Curiosities of the Plant Kingdom

Reinhardt Höhn
in collaboration with Dr. Johannes Petermann

Curiosities
of the
Plant
Kingdom

CASSELL
London

CASSELL LTD.
35 Red Lion Square, London WC1R 48G
and at Sydney, Auckland, Toronto, Johannesburg,
an affiliate of
Macmillan Publishing Co., Inc.,
New York

Translated from the German by Herbert Liebscher
Copyright © 1980 by Edition Leipzig

First published in Great Britain 1980

Reproductions in the text, if not noted to the
contrary, from: Leonhard Fuchs, *Newes Kreüterbuch.*
Basle, 1543

Designed by Walter Schiller, Altenburg
Drawings by Klaus Bethge, Berlin
Production by Graphische Werke Zwickau
Printed in the German Democratic Republic

ISBN 0 304 30463 8

Table of Contents

Foreword   6

Where curious plants grow   7

The tropical rain forest   9
Deserts and semideserts   12
The Mediterranean region   33
The Canary Islands—a plant museum   34
Peculiar Cape flora   35
Primeval Madagascar   36
Sclerophyllous vegetation in Australia   37
On the boundaries of vegetation   39

Useful beauty   57

On the lookout for pollinators   59
Trapped pollinators   60
Diversity of orchids   63
Fooled males   64
Complicated forms of fertilization   65
Birds and bats   66
Opening times of various flowers   68

Specialists in the plant kingdom   93

The struggle for light   95
In marsh and swamp   97
Halophytes   100
Carnivorous plants   101
Viviparous plants   105
Plants without roots   106
Plants without chlorophyll   107

Magic plants—witch plants   133

Plants and superstition   135
The wonder-working mandrake   136
The gist of the matter   138
Drugs and poisons   139
From arrow poison to drug   141
Ginseng between superstition and reality   142
The road to "Paradise"   143

The fossil book of nature   153

The first traces of life   154
In the Carboniferous forest   155
Seed formation   157
A strange discovery   159
The great change   160

About giant and millenarian plants
and other curiosities   169

Giant trees   171
Giant growth of other plants   173
Millenarian plants   175
Strange shapes, rare plants   176

Appendix   195

Glossary   196
Index of plant names   207
Bibliography   211
Sources of illustrations   212

Foreword

After Alexander von Humboldt, in 1804, had returned from his extended journey through the equinoctial regions of the New World and published his experiences under the title *Aequinoctialgegenden des Neuen Continents*, his contemporaries were as fascinated and enthralled as von Humboldt himself by his *Natural Painting of the Tropical Countries* (1807), by "cactic forms," "barrel trees," "mimosa ferns," and many other things which he had seen during his journeys and that were unknown in Europe at that time. Already, Humboldt found that "all animal and vegetable development is bound to fixed, eternally acting types which determine the physiognomy of nature." For this reason, the author starts his book with descriptions of the habitats of "rare plants," the tropical rain forest, deserts and semideserts, primeval Madagascar, and the boundary regions of plant life.

Although the author deals with giant trees, millennial plants, and other curiosities, he is not so much concerned to present, in word and picture, odd shapes and formations of plants that will strike the eye of the uninformed and nonbotanists, but he is much more anxious to point out those less conspicuous peculiarities that the layman will normally miss if his attention is not drawn to these items. Thus, the author presents in word and picture carefully selected examples of interesting pollination mechanisms, opening times of various flowers, Linné's floral clock, humming birds and bats as pollinators, ecologically and physiologically specialized groups of plants such as epiphytes, halophytes, rootless plants, parasites, and saprophytes as well as magic plants, witches' plants, and poisonous plants. Finally, the author deals with the "fossil book of nature," showing that the present plant cover of the earth is the result of a development of many millions of years. Wherever you open the book, you will find "curiosities of the plant kingdom."

Text and illustrations will appeal not only to the expert but also to the amateur botanist, giving him valuable information for his roving expeditions in nature so that he may experience for himself all of those curiosities.

Werner Rauh, Heidelberg

The tropical rain forest

Deserts and semideserts

The Mediterranean region

The Canary Islands—a plant museum

Peculiar Cape flora

Primeval Madagascar

Sclerophyllous vegetation in Australia

On the boundaries of vegetation

Where curious plants grow

When the traveler from Europe first sets foot in the forests of South America, he is confronted with a completely unexpected scene. Everything he sees is only faintly reminiscent of the descriptions essayed by renowned writers on the banks of the Mississippi, in Florida, and elsewhere in the temperate climes of the New World. At every step he feels that he is not on the threshold of the torrid zone but in its midst, not on an island of the Antilles but in a mighty continent, where all is of gigantic proportions: mountains, rivers, and vegetation. If he appreciates natural beauty, he scarce knows how to order his varied impressions. He is unable to say what fills him with greater awe, the majestic stillness of solitude or the beauty of the individual forms and their contrasts or the vigor and luxuriance of the vegetable world. It seems to him as if the soil has no more room for the development of the overflowing abundance it nourishes. Everywhere the trees are cloaked in a carpet of green. To take all the orchids, all the species of pepper and anthurium that grow on a single locust tree or on a fig tree *(Ficus gigantea)* and plant them all out would be to cover a fair amount of ground. By these wonderful accumulations, forest, rockface, and mountainside extend the realm of organic nature. The same lianas that trail along the ground attain to the tops of trees and, more than thirty meters up, cross from one to the other. Parasitic growths are present in such profuse confusion that the botanist is ever in danger of mixing blossoms, fruits, and leaves that belong to different species. We proceeded for some hours in the shade of these green vaults, through which one rarely espies a glimpse of the blue sky. A narrow path leads, winding, from the edge of the forest through open, singularly moist country. In temperate climes grasses and reeds would have been growing on such land, in a wide expanse of meadow. Here the soil is lost in a profusion of aquatic plants with arrow-shaped leaves, in particular species of canna. These succulent growths attain a height of two or three meters and, close together, would pass in Europe for little woods. This magnificent scene of meadowland and a grassy carpet interweaved with blooms is totally alien to the lower stretches of the torrid zone, and is found again only on the Andean plateaus.

This impressive description of nature was written at the beginning of the 19th century. Its author, the famous scientist Alexander von Humboldt (1769–1859), was one of the founders of plant geography. In his thirty-volume work *Reise in die Aequinoctialgegenden des neuen Continents*, which includes contributions by other distinguished scientists such as Aimé Bonpland, Georges Cuvier, Valenciennes, Oltmanns and Kunth, he published comprehensive material on South and Central America with detailed descriptions of the native floras. In several other works, including *Ideen zu einer Geographie der Pflanzen* (1807), Humboldt tried not only to show his readers the flora and vegetation of foreign countries but also to explain why different floras existed in various regions. At that time, he met with a responsive and astonished public. Today it is considered a matter of course that there are different plants in the various geographical latitudes, that tropical lianas do not grow in the Sahara Desert and that firs are not to be found in tropical rain forests. But again and again we are amazed at the great diversity of plants encountered in various regions. For this reason we think it useful to look more closely at a few interesting zones of plant geography at the beginning of this book.

The most exuberant vegetation of the earth is to be found in the tropical rain forest (Ill. 1). This hot, damp vegetational area is restricted mainly to the equatorial region, which reaches from 10° S. to 10° N. The tropical rain forest in South and Central America extends to a large continuous region from the Amazon Basin to the east side of the Andes; it is found in the Indo-Malayan region; in West Africa in the Congo Basin and the coastal regions of the Gulf of Guinea; and in smaller, scattered areas on the rainy hillsides of eastern Brazil, north-eastern Australia, India and East Africa. In all of these regions, beside extremely heavy rainfalls, uniformly high temperatures prevail all the year round. The temperature never drops below 18 °C. It varies slightly according to the time of day but not the season of the year. In some areas, peak precipitation exceeds 10,000 mm per year but often is considerably lower. It is, however, never as low as in Central Europe, where mean precipitation varies from 500 to 1,000 mm in the individual regions. Heat and moisture have contributed toward the development of a vegetation that is characterized by a wealth of lofty trees, among which giant trees 40 to 50 m in height are not rare. In growth, shape and pattern of life, however, these trees are clearly different from those found in the forests of the temperate zones. Striking features are the relatively small tops and large, firm, frequently leather-like evergreen leaves, which in most cases have so-called dripping tips. We never see the tropical rain forest leafless, nor do we see it fresh and green, a sight found in early spring in the deciduous woods of the northern hemisphere. The forest is always uniformly green because the development of new leaves is independent of the time of year and therefore is irregular. Because of the lack of pronounced seasonal variations in temperature and precipitation, the growth of the trees is uniform in height and thickness throughout the year, so that no annual rings are formed in their trunks, as is the case with trees growing in temperate zones.

In the soils of the tropical rain forest, nutrients are concentrated in the upper layers of the ground alone. For this reason, even the tallest trees put forth relatively flat root systems which, however, do not give them a firm hold. The

Hibiscus

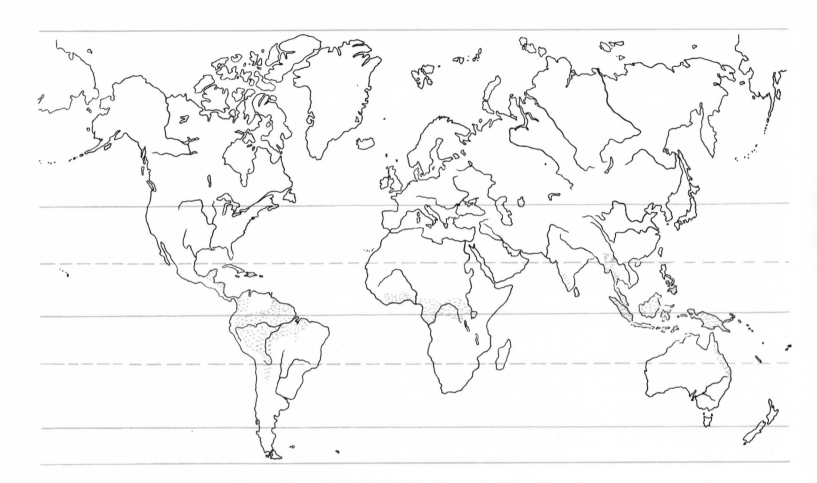

trees adapt themselves to these environmental conditions and frequently produce huge board-like roots, as can be seen in *Coussapoa dealbata* (Ill. 2).

Pollination is carried out almost exclusively by insects, birds, and bats, because, with the permanent foliage, pollination by wind would hardly stand a chance. For this reason, the blossoms are frequently large, showing a brilliant coloration. With a great number of tropical trees, the buds lie, almost until opening, in the closed calyxes in water, as exemplified by the West African tulip tree *(Spathodea campanulata)* (Ill. 3). The reasons for this phenomenon are not yet fully understood.

With some trees, the flowers appear on the leafless trunk and on old branches. This botanical peculiarity, which is called cauliflory, is associated with the pollination of flowers by bats and with the distribution of seeds by flying foxes. The animals get to the trunks of tropical trees more easily than into the dense tops. A typical example of cauliflory is the cocoa tree *(Theobroma cacao)*, with its small rose-colored flowers (Ill. 4). Each year a mature cocoa tree produces up to 100,000 flowers, but only a fraction of these—about 5 per cent—get pollinated. Because of shedding of flowers and the premature deaths of young fruits, the proportion of ripe fruits drops as low as 0.5 per cent. In general, the fruits are called siliques. Botanically, however, they are berries that contain 20 to 40 seeds, each of which is surrounded by a sweet-sourish pulp. The fruit reaches a length of up to 30 cm, has a diameter of 5 to 10 cm, and may have a weight of up to 500 g.

After they are harvested by man, the seeds, the cocoa beans, which are rich in protein and oil, are fermented, dried, hulled, and pulverized. Fat, which is also known as cocoa butter, is pressed out of the cocoa mass, and the remains are marketed in the form of cocoa powder. The cocoa powder contains the stimulating alkaloid theobromine. In some regions, cocoa beans are even used as currency. Although the tops of the trees in tropical rain forests are relatively small, they form a rather dense roof, so that only a small quantity of light can penetrate. This has a bearing on the development of the undergrowth. In these forests, the undergrowth consists primarily of young trees and high to very high shrubs. The poor light conditions exert an even greater influence on the ground stratum of growth than on the undergrowth stratum; in the former, green plants are found only in places where sufficient light is cast on the ground. Some plants have adapted themselves to the low light intensity in that they have reduced breathing, thus possessing a balanced metabolism in which reduced anabolism and reduced catabolism are in equilibrium. Examples are indoor plants such as ornamental vines *(Cissus)*, which can thrive even in relatively dark rooms.

Frequently, fungi as well as parasitically and saprophytically living higher plants grow in the ground stratum of tropical rain forests. An example of a particular adaptation to the light conditions prevailing in the forests are a great number of lianas and epiphytes, which are discussed in greater detail elsewhere in this book. Another striking feature of tropical rain forests is the great wealth of plant species. For example, more than one hundred different species have been found on an area of one hectare.

Notwithstanding this abundance of species, certain species or plant groups are characteristic of the individual geographical regions of the tropical rain forest. In the South American forests these are the Pará rubber tree *(Hevea brasiliensis)*, the Brazil nut tree *(Bertholletia excelsa)*, the "royal" water lily *Victoria amazonica* as well as many bromelias, and a large number of orchids, including the showy, fragrant genus *Cattleya*, which is frequently grown in greenhouses.

Typical plants of the African rain forest are the oil palm *(Elaeis guineensis)* and the caoutchouc liana *(Landolphia)*, also known as Congo rubber. The Asian regions of the tropical rain forest include as characteristic species representatives of the sago palms (genus *Metroxylon*), the rotan palm *(Calamus)*, which is discussed further below, and typical genera of orchids, among them *Dendrobium* and *Canna*.

Deserts and semideserts

Deserts and semideserts grossly differ from the tropical rain forests. High temperatures in conjunction with small amounts of rain throughout the year have led to the formation of peculiar shapes of plants in these regions of the subtropical zone. These arid regions are of a quite distinct nature in the tropics. In the semideserts, vegetation is so sparse that most of the ground surface is not covered by plants. In the deserts proper, the ground surface is almost completely without vegetation over a large part of the year. Rainfall is irregular, reaching less than 200 mm per year. Because of the extremely scarce formation of clouds, the differences in temperature between day and night are very pronounced. Temperature variations of 40°–50 °C are not out of the ordinary.

The plants have adapted themselves to these extreme environmental conditions by producing a great variety of special forms. Some of the plants are rather short-lived, and they avail themselves, during the short periods in which water is present, of the opportunity for growth, flowering, and seed formation. The seeds remain in the soil, capable of germination, for prolonged periods. Other plants survive the long droughts as tubers or bulbs in the soil, and there are also shrubs that grow hardly any leaves.

The most striking plants in the arid regions, however, are the succulents. These plants are capable of storing water in leaves, stalks, or roots. In this way they can survive prolonged spells of dry weather. There are trunk succulents, which have no leaves. Such plants, many of which have stood motionless in the dreary and dry areas for centuries, impart an impression of primeval eras. Among the best-known succulents are cacti, which occur mainly in Central and South America, and the spurges (Euphorbiaceae), as well as the sedums (Crassulaceae), fleshy herbs that grow mainly in Africa. The largest desert in the world is the Sahara in North Africa, which has an area of approximately 9 million square km. Annual precipitation is less than 50 mm. Vegetation is scarce, and vast areas are covered by nothing but sand. Oases exist only in places where water supply is furnished by local seepage or springs. The most important tree is the date palm *(Phoenix dactylifera)*, which is used as food for man and animal and for building material (Ill. 5).

The Namib Desert on the coast of South-West Africa is also an extremely arid region. There and in the neighboring areas of the Kalahari Desert and the Great and Little Karroo plateaus of a semidesert character, very strange plants grow. The most renowned plant of the Namib Desert is the welwitschia *(Welwitschia mirabilis)* (Ill. 6). This odd plant may impart the impression of a representative of antediluvian floras. The thick trunk penetrates deep into the soil. The woody overground part resembles a large funnel in shape and has a split rim. Since the trunk is less than a foot high but large in diameter, it has the appearance of a disk; it is covered by cork and soaks up rainwater like a sponge. The plant produces only two persistent, leather-like, blue-green leaves that grow at the base and die at the apex. The rod-shaped root grows 1 to 1.5 m in length and draws water from the deeper layers of the soil whereas a side root system immediately takes up any water from the rare rainfalls that wet the upper soil layers.

Welwitschia mirabilis is of special interest not only because of its particular capability of adapting to environmental conditions but also because it is virtually isolated in the plant kingdom. It is the remainder of a developmental branch that morphologically is positioned between the gymnosperms and the angiosperms. Today Welwitschia grows only on a fairly small tract of the Namib Desert extending at a distance of 50 km from the shore. The plants may reach an age of more than one hundred years. A particularly large specimen was found to be over 500 years old. This longevity is one of the factors that have prevented Welwitschia from dying out. Because of their particular systematic position and the isolated geographical distribution, the few existing specimens have been protected by strict nature conservancy regulations since the beginning of the 20th century.

The naras *(Acanthosicyos horridus)*, spiny shrubs with a fruit that resembles a melon, cover the sand hills in the Namib Desert with a tangled growth of long sprouts that have thorns but no leaves (Ill. 8). Roots up to 15 m in length and as thick as a man's arm provide the plant with water. Succulents of different forms also grow in the Namib. Besides tree-like species of aloe *(Aloe dichotoma, Aloe pillandsii)*, monstrous

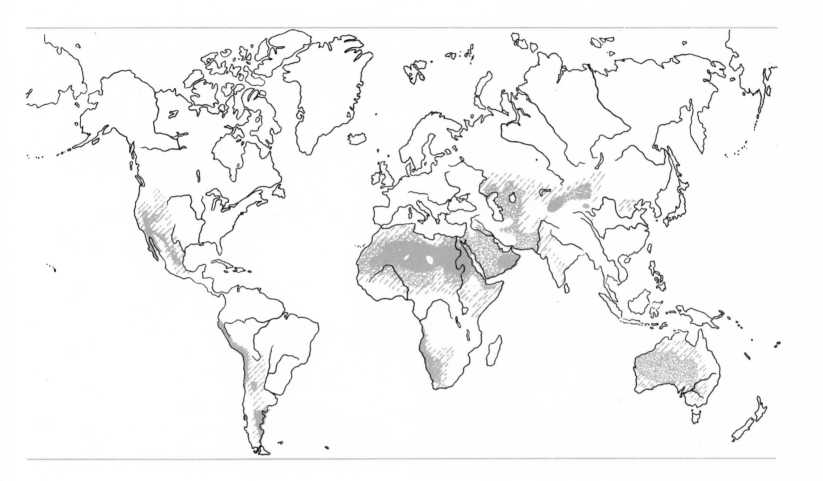

trunk succulents of the genus *Pachypodium* are found. *Pachypodium namaquanum* is a characteristic plant of rocky quartzite slopes (Ill. 11). The specimens form 2-to-3-m-high columns that bear a small cluster of hairy and decidedly undulated leaves on their top ends. This leafy crown is always directed to the north, a peculiar feature of this plant. Reddish-brown, velvet-like flowers grow between the leaves. The arid areas in southern Africa, where the smallest amount of rain falls, are populated by the highly succulent *Hoodia bainii* (Ill. 7). This bushy, richly branching shrublet grows to be 10 to 40 cm high, putting forth gray-green, thoroughly ribbed shoots. The showy flowers are bowl shaped and very large, varying in color from yellow to red and purple.

Other succulents of a striking appearance are members of the genus *Huernia*. The flowers of *Huernia zebrina* (Ill. 9), which grow on the sands of the South African desert, emit the smell of carrion. Frequently the flowers are larger than the whole plant. The pronouncedly concave ring in the center of the flower resembles a chocolate bar. The brimstone-colored flowers of *Huernia hystrix*, with their red spots and stripes, attract particular attention (Ill. 10). Large numbers of fleshy and spine-like papillas grow on the whole plant, so that it gives the impression of hedgehog. The greatest marvel in these arid regions are the "flowering stones." These mostly spherical or angular stone-like plants, which frequently consist of only one pair of leaves, are unique in the plant kingdom. These wondrous plants grow between round pebbles and angular pieces of rocks. Color and shape are adapted to the environment, so that they cannot be distinguished from the surrounding stones (Ill. 12). It is only when they open their beautiful and gorgeous flowers that they are set off against the stony ground (Ill. 13). This camouflage for protection from hungry animals—known as mimicry—has developed in the course of millennia. The "living stones" belong to various genera, primarily to the *Lithops* species; but there are also plants of the genera *Conophytum, Gibbaeum, Fenestraria, Pleiospilos,* and *Titanopsis,* which constitute this mimicry group (Ill. 14).

On their surface, these plants have dot-like eyes that act as windows. These are light wells that permit light to fall deep into the water tissue. They enable the plant to assimilate over a large area of its mass, although its external surface is very small. The flowers usually are of extraordinary beauty. Deep and shining in color, they generally cover the whole body of the plant. Often the smallest plants possess the largest flowers. The fruits, which are divided into individual chambers and get lignified, are true masterpieces of nature. With the onset of the rains, the lids of the chambers are opened by a swelling device, and the raindrops bounding against the fruit throw the seed kernels out of the chambers onto the desert sand.

The arid areas in East Africa are characterized by quite different plants. Peculiar succulents grow in what is probably the driest region near the equator, the area in East Africa between the Pare and Usambara Mountains, where the annual mean temperature is 28 °C, the soil temperature at a depth of 40 cm is 30 °C, and the annual rainfall is between 100 and 200 mm.

The desert rose *(Adenium obesum)* produces large pink flowers in the leafless state. The thick fleshy stem gets twisted as it grows and has a misshapen appearance when old. The milky juice is used for producing a strong arrow poison.

The dry savannas are covered with xerophytic grasses, among which the high elephant grass *(Pennisetum benthami)* is dominant. In these regions real trees also occur—e.g. the umbrella acacia *(Acacia spirocarpa)* and the tamarind *(Tamarindus indica).* The baobabs *(Adansonia digitata),* with their enormous and thick trunks, are the giants among the plants. These trees are the characteristic plants of the East African thorny savanna.

The fruits of the sausage tree *(Kigelia pinnata),* which occurs in the East African thorny savanna, are suspended on long stalks (Ill. 19). The flowers appearing on these long shoots are reached by bats, which pollinate them much more easily than they do flowers concealed in dense treetops.

Cacti showing many shapes and forms are characteristic of the desert regions in North, Central, and South America. They grow in the form of cushions having a height of hardly 5 cm, big spheres, or lofty columns rising far into the blue skies (Ill. 16).

When talking of giant cacti, saguaros *(Carnegia gigantea)* are invariably mentioned. These are the tallest cacti, at least in the United States (Ill. 20). These unusually imposing plants grow in a large nature preserve, where they are protected by stringent regulations to prevent their decimation or eradication. Their natural habitat comprises extensive semideserts in southern Arizona and southeastern California, as well as in the Sonora Desert in northern Mexico. They reach a height of up to 12 m, and their trunks grow to a diameter of up to 65 cm. In advanced stages of growth, they branch out in the form of a candelabrum; that is why they are also called candelabrum cacti. Their parallel rising branches, usually five or six in number, start growing only when the main stem has reached a height of 5 to 8 m. The age of these giant specimens, having a weight of anything between six and eight tons, reaches hundreds of years, for their height increases by one meter every 30 to 50 years, as was found by the desert laboratory in Tucson in the course of studies conducted over several decades. The cacti become reproductive only when they have reached a height of several meters. Numerous flowers appear at the apex of the main stem and each branch. Several species of bees perform pollination. The fruits are 6 to 9 cm long, filled with red pulp and blackish seeds when fully mature. They are collected and eaten raw or dried by the Indian tribe of the Papago, whose reservation is situated within the nature preserve. Most of the fruits, however, are eaten by birds. One plant can produce up to two hundred fruits a year, each containing about one thousand seeds; they become ripe within two to four weeks before the summer rains, that is to say, before the main growth period, which lasts from July to September. Strikingly characteristic of *Carnegia gigantea* are its crooked branches; the crookedness is produced by drooping (during drought) and subsequent erecting (after rainfall), so that the plant assumes a queer shape. Also, there may be crest-like connections between several branches, so-called cristated forms, but these appear only in very old specimens. Another interesting feature is the fact that the roots of the cacti do not penetrate deep into the ground. They grow laterally to a considerable extent, so that they can take up large quantities of rainwater. The side roots of *Carnegia gigantea* extend up to 30 m in all directions. As soon as the soil is wetted, they form fine sucking roots, within a period of 24 hours, which draw in water. They are capable of storing 2,000 to 3,000 liters of water. Without suffering any damage, the plants will then survive without any precipitation for as long as a year.

The hatchet-shaped, closely spaced warts, which are densely covered by thorns in a comb-like arrangement, give *Pelecyphora aselliformis* a very strange structure (Ill. 18). The initially one-headed plant develops a dense white wool between the warts and produces a large number of carmine-violet flowers from the woolly crown. This interesting cactus is an indigenous plant of northern Mexico, where it is found in dry, sunny places. Later, branching out by lateral sprouting, smaller groups are formed; these offshoots, however, rarely exceed a height of 10 m.

A unique and very striking specimen of the tall columnar cacti of Mexico is the old-man cactus *(Cephalocereus senilis)* (Ill. 17). It has become known throughout the world under the name *cabeza de viejo* (Spanish), which was coined by the Mexican aborigines, or an equivalent term in other languages. This attractive plant, with its long, white, curly hair, enjoys great popularity among cacti fans. Its natural habitat in the hot valleys in central Mexico is characterized by a dry, hot climate which, however, changes to a damp, hot condition during the vegetation period. Normally, the enormous and imposing columns become capable of flowering only when they have reached a height of 6 to 8 m. Their stems have a diameter of up to 30 cm, and they develop a peculiar crown, called cephalium, which in its initial stage grows on one side only and later will gradually grow all around the top of the columnar stem, seldom remaining on one side only. In this flower region, hair is considerably shorter, interspersed with shaggy wool, giving the impression that the end of the stem is wrapped in lambskin. Twenty to thirty straight ribs of a stem, which are only about 3 mm deep, are rather densely covered with areoles bearing perfectly white tufts of hair that loosely envelop the plant body. The wool of the cephalium protects the developing bud and covers the fruits until they are ripe, whereas the nocturnal flowers peep out of this cover; they are funnel-shaped, about 5 cm long and 7 cm wide, and whitish-pink in color. Although old specimens reach a height of about 15 m, the woody tissue developed in the interior of their stems is negligible and thus their anatomic structure shows a striking difference to that of many other columnar cacti, which yield timber to an appreciable extent for the native population. This also explains the fact that strong winds sway the gigantic columns, frequently breaking them.

The first cacti which were encountered by the Europeans on their arrival in the New World were species of the genus *Melocactus*. The strange plants native to the coastal regions naturally attracted the attention of the Spanish newcomers who had never seen such plants before. While the vegetative parts of the plants, which resembled thick cucumbers or melons, appeared to them very curious, they were even more amazed at the cephalium with usually a large number of reddish flowers and bright red fruits.

The cephalium—the section of the shoot where flowers are formed—in a *Melocactus* initially is small, flat, and round but will become cylindrical in shape during its further development. Every year it gains in height; in some species it may grow 20 cm high and, under particularly favorable conditions, even higher.

Melocactus acunai (Ill. 21), which occurs on the coast of Oriente Province in Cuba, is one of the species whose melon-shaped body frequently is crowned by branched or divided cephalia. Further they possess the capability of putting forth new shoots from their tops, but these remain small as compared with the parent body, although they, too, will soon form new cephalia.

Ritterocereus hystrix, a cactus that may become more than 8 m high, also occurs near the shore in Cuba and on the neighboring isles (Ill. 21).

On the whole, we can say that quite different cacti abound in the Sonora Desert of northern Mexico and Lower California, as well as in the debris deserts of Mexico and the stone desert in northern Peru (Ill. 22). Beside cacti there are other plants of an aboriginal appearance growing among heaps of boulders, such as the yuccas which have high trunks that bear a head of rigid leaves extending into the sky and a cloak of the dead leaves hanging down close to the trunk.

Because of cold ocean currents, foggy subtropical deserts have developed in the western coastal region of Peru and Chile. The soil receives water only through the fog, because this region is almost rainless. Among flowering plants here, a few tillandsia species (e.g. *Tillandsia purpurea*), which live only on fog water (Ill. 23), grow as the only fog plants known in the world. The small rosettes of these bromeliads bear many tiny scales which take up condensed water. The plants are loosely seated on the sandy soil; they do not possess roots. Plants other than these small-leaved bromeliads cannot live in the foggy desert.

In Central Asia, too, there are vast deserts and semideserts. Vegetation is scarce; large areas are sandy, dreary, and vegetationless. Plants have adapted to the extreme conditions in a great variety of forms.

Saxaul plants, species of the genus *Haloxylon*, which grow in Central Asian sandy deserts, are called "trees of the desert." They have a bushy growth and their leafless branches are their organs of assimilation. They retain blown sand, endure extreme droughts, and tolerate the high salt content of the soil. They represent the only source of wood that can be used for commercial purposes.

While the black saxaul *(Haloxylon aphyllum)* thrives on the saline clayey soils in Central Asia and in the Kazakh S.S.R., the white saxaul *(Haloxylon persicum)* (Ill. 24) grows in the gypseous stone deserts of the Kara Kum and Kyzyl Kum. By afforesting regions in Central Asia that are destitute of forests, using the black saxaul, the population is supplied with fuel. At the same time, this method serves to stabilize shifting sand dunes and exerts a favorable influence on the development of pasture plants so that karakul husbandry can be expanded. This species of saxaul is also used as fodder.

3 Inflorescence of the African tulip tree *(Spathodea campanulata)*
4 Flowers and fruit of the cocoa tree *(Theobroma cacao)* developed
directly at the stem

5 The date palm *(Phoenix dactylifera)* is the characteristic tree of the Sahara.
6 One of the earth's most renowned and rare plants is *Welwitschia mirabilis*, which occurs in the Namib Desert of South-West Africa.

7 The highly succulent *Hoodia bainii* grows in regions of South Africa with the smallest amount of rainfall.
8 The nara pumpkin *(Acanthosicyos horridus)*, having leafless but long shoots with green thorns, also grows in the Namib Desert.

9 The flowers of *Huernia zebrina* lying on desert sand smell intensely of carrion.
10 *Huernia hystrix* of South Africa bears strikingly beautiful flowers.

11 *Pachypodium namaquanum* grows on slopes of quartzite
in South-West Africa's Namaqualand.
12 The non-flowering plants of *Lithops meyeri* can hardly
be distinguished from stones.
13 Flowering plant of *Lithops aucampiae*

14 *Pleiospilos bolusii* is an example of the so-called living stones of South Africa's desert regions.

15 Cacti of the species *Thephrocactus floccosus* which grow
in the upper regions of the Andes are called "living snow."

17 The old-man cactus *(Cephalocereus senilis)* is a striking Mexican type.
18 The thorny pads of *Pelecyphora aselliformis* are reminiscent of isopods.

20 Saguaro cacti *(Carnegia gigantea),* sometimes gigantic in size, produce bizarre shapes in the American desert.
21 *Melocactus acunai* (foreground) and *Ritterocereus hystrix* (middleground) by the shore of Oriente Province in Cuba

23 *Tillandsia purpurea* is a typical fog plant along the coast of Peru and Chile.

24 The Saxaul species (genus *Haloxylon*)
are characteristic plants of the Central Asian desert.

Transition from the subtropics to the temperate zone is not uniform according to latitudes. Interlocking of land and sea is characteristic of this region. Precipitation fails to show a distinct annual distribution. Temperature, however, is an important feature of this region because of a more or less prolonged colder season of the year. Vegetation in the Mediterranean region is clearly characterized by the prevalence of sclerophyllous copses. Similar regions are found on the Pacific coast of North America, in central and southern Chile, and in parts of South Australia.

Here we shall deal exclusively with the Mediterranean region, especially because its vegetation has been strongly influenced by man to a greater degree than anywhere else on earth. With regard to plant geography, this region is almost identical to the original area of distribution of the European oil palm *(Olea europaea)*.

Before man removed or badly affected the natural vegetation in the Mediterranean region by destroying forests and forming extensive pastures, this region was covered by evergreen deciduous woods, where holly oaks *(Quercus ilex)* prevailed. Apart from the oil palm, other important types of timber were the cork oak *(Quercus suber)*, which grew in the western part, and the Kermes oak *(Quercus coccifera)* in the eastern part, as well as the carob tree *(Ceratonia siliqua)*.

Because of the action of man, forests were decimated, large tracts of fertile soil were eroded, and hillsides were reduced to exposed bare rock. Today, the characteristic tree of the Mediterranean region is the stone pine *(Pinus pinea)*, which has a widespread, flat-topped head (Ill. 25), but cluster pine *(Pinus pinaster)* and Aleppo pine *(Pinus halepensis)* also are important, and all three species are used in reforestation.

Maquis, a thick scrubby underbrush, evergreen, almost impenetrable, and as tall as a man, has developed into a characteristic vegetation type in the Mediterranean region. The above mentioned oak species grow as shrubs among maquis. Characteristic is the strawberry tree *(Arbutus unedo)* (Ill. 26), an evergreen with dense foliage, which grows up to 10 m and puts forth scarlet fruits resembling strawberries, which are edible but not savory. The Latin name *unedo* (= I eat one) means that one

berry is sufficient. In the eastern part of the Mediterranean region, *Arbutus unedo* is replaced by *Arbutus andrachne.* Other plants of the maquis are well-known species such as the mastic tree *(Pistacia lentiscus)*, the myrtle *(Myrtus communis)*, the tree heath *(Erica arborea)*, and the oleander *(Nerium oleander).*

In the course of a further worsening of the living conditions of plants, the maquis gradually changes over to heaths and grass communities which are called *garigue* in France and Italy, *tomillares* in Spain, and *phrygana* in the Balkan Peninsula. In these plant communities we may find, depending on the soil conditions, lavender *(Lavandula stoechas)*, rosemary *(Rosmarinus officinalis)*, wild tulips *(Tulipa sylvestris)*, grape hyacinths *(Muscari neglectum)*, daffodils *(Narcissus poeticus)*, rockroses *(Cistus salviaefolius* and *Cistus monspeliensis)* and other species that are also known in many gardens.

Some orchids also grow in these plant communities. Inhabitants of open grassy or stone slopes, preferably on a chalky underground, are species of the insect orchis *(Ophrys)* (Ill. 27, 28). A particular feature of this genus of orchids, which comprises about thirty species and is typical of the Mediterranean region, is the shape of the flower lip, which largely resembles different insects, depending on the species. The biologic peculiarities associated with this are dealt with later in the text.

If one glances at the Serapias species (Ill. 29, 30) one will not readily recognize the typical orchid flower. The habitats of these plants with their queer flowers are damp meadows, dry, grassy places in the cistus garigues, olive groves, and evergreen forests of the Mediterranean region. The garigue, especially, a plant community loosely spread over dry, rocky areas, forms the gay Mediterranean spring landscape with these orchids. After the bursting into bloom of many plants from March to May, the dried-up landscape typical of the Mediterranean region appears rather suddenly.

Mention should also be made of a particular plant of the Mediterranean region: the dwarf palm *(Chamaerops humilis)*, the only palm that grows wild in Europe (Ill. 31). Many parks and gardens cultivate it. This low-growing plant, which rarely forms a stem, develops many flowers of a dioecious distribution. The dwarf palm can stand temperatures below 0 °C.

The Canary Islands—a plant museum

The flora of the Canary Islands may be compared with a museum. There specimens of plant species have survived that died out in South and Central Europe during the glacial period. In addition, representatives of the South African flora—tropical and subtropical plants—as well as plants that originally grew in the Sahara Desert and in Alpine regions remain in existence.

In the volcanic island of Tenerife, where the 3,718-m-high Pico de Teyde is the greatest elevation, there are different vegetational units because of the different altitude levels. Deserts, semideserts of succulents, and laurel woods, pine forests, mountainous deserts, and Alpine slopes of rocky debris show very interesting and rare plants to the looker-on. Unfortunately the original vegetation has been changed by man here, too, so that only remnants still exist.

The landmark of the Canary Islands is the dragon tree *(Dracaena draco)*, which may grow up to 18 m high and very stout (Ill. 33). Paleobotanic finds have shown that its shape has remained unchanged since the Eocene epoch—i.e. for 60 million years. It belongs to the lily family and is one of the few trees among the species of this group. It is closely related to the bowstring hemp of *Sansevieria*, which is well known as an indoor plant. When the trunk is injured, the bark of the dragon tree emits a reddish juice, which is said to turn into blood, namely dragon's blood, possessing magic capabilities. Formerly, the resin was used for medicinal purposes, but today it is used in the manufacture of varnishes and polishes. To the botanist, the dragon tree is of interest also because of its particular secondary growth in thickness.

Other trees characteristic of the Canary Islands are the Canary Island date palm *(Phoenix canariensis)* and the Canary Island laurel tree *(Laurus azorica)*.

On young and dry lava soils, a white-gray species of spurge having the shape of a cactus *(Euphorbia canariensis)* is predominant; it starts branching out directly from above the ground, covering large tracts (Ill. 34). Heights of 6 to 8 m are found fairly often. The plants contain a poisonous milky sap that produces a pungent and caustic effect. The Roman writer Pliny the Elder (23–79) already mentioned this plant.

One of the rarest plants in the world is the red viper's bugloss *(Echium wildpretii)*, which grows only on a small area of about 1 hectare in the mountains encircling the Pico de Teyde on Tenerife at an altitude of 2,200 metres (Ill. 32).

From a rosette of long and narrow leaves, a red, racemose inflorescence, 2 to 4 m long, rises skyward. The flowers are borne on their stalks along an unusually long shaft twisted in the form of a left-hand spiral. The plant requires several years until it can bloom; subsequently it dies.

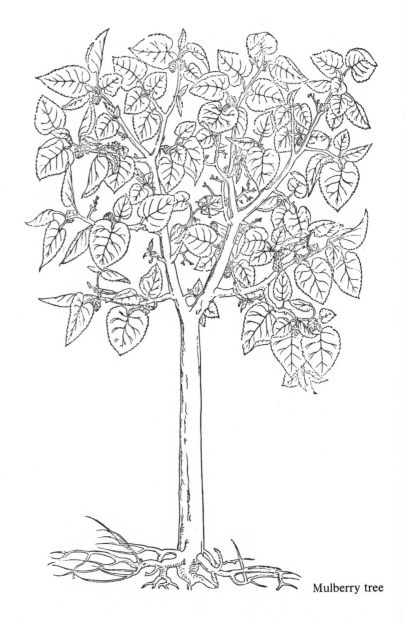

Mulberry tree

At the extreme southwest end of Africa, a sclerophyllous vegetation has survived in a relatively small area that is of particular interest to the botanist not only because of its unusual abundance of species and floristic peculiarities but also because of the history of the floras. Many scientists consider the small area of Capensis, which is only about 50 km long and up to 10 km wide, as the special floral region. On the mainland it is bounded by the mountains of the Karroo, on the other side of which the South African semidesert region is situated. In the Capensis region more than 6,000 different species of flowering plants have been counted. To form an idea of the significance of this number, one should realize that only about half that number of plant species are native to the entire area of Central Europe.

The prevalent vegetational forms of the Cape region are heaths and shrubbery where, as a special feature, shrublike plants of the composite family occur—e.g. the genera *Osteospermum* and *Erytropappus*. The presence of a great number of heather plants and *Olea verrucosa*, a relative of the olive tree, suggests that there are some phytogeographical relations with the Mediterranean region.

The characteristic tree of the Cape region is the silver tree *(Leucadendron argenteum)*, which, however, occurs on the damp slopes of the Table Mountain in a very small area below an altitude of 500 m (Ill. 35). This tree, with a dense foliage, grows up to 15 m high and has long silvery, silky leaves covered with hair. Every hair has a basal cell functioning as a joint enabling the hair to move. When there is a sufficient supply of water, the hairs form an angle of about 30 degrees with the leaf surface so that air can easily get to the stomata. In the event of water shortage, the hairs bend over, one upon the other, forming an effective seal, thus restricting the gaseous interchange between the atmosphere and the intercellular spaces within the leaf, and limiting evaporation. Therefore, the trees show a brighter luster in summer. In sunshine the trees seem to be covered with silk. Yellow flower heads appear at the ends of twigs, with the yellowish spathes widely spread and extending beyond them. The silver tree is a plant of the family Proteaceae, which is composed of 1,200 species that occur in the southern hemisphere mainly in Australia and South Africa.

The genus *Protea*, also native to the Cape region, comprises both shrubs of 10 m in height and dwarf plants that show nothing but their leaves and flower heads above the ground (Ill. 36). The leaves, including many very large ones, have a thick cuticula and, thus, are very hard, and some of them are hairy, like the leaves of the silver tree. The wax film in *Protea grandiflora* is so thick that the leaves shine like white balls. The basket-shaped inflorescences are at the ends of twigs and surrounded by brilliantly colored bracts. The striking colors of the flowers and the nectar at the bottom of the calyces attract honey eaters, which perform pollination.

Individual species of the genus *Protea* may grow very old. Frequently, plants are destroyed by bush fire. Since they possess very resistant seeds, new plants will soon start growing again. Only *Protea grandiflora* is known to survive fires because of its subterranean parts, which will put forth new sprouts.

The flowers of *Protea barbigera* (Ill. 38), many of which can reach a length of 25 cm, resemble mature pineapples. The whole flower head is covered with white down borne on the yellowish bracts which show blood-red dots at their tips.

Artichoke-like flowers are formed by *Protea cynaroides* (Ill. 40). The flowers, of a pale pink color, have white, tomentose, involucral leaves and are borne on reddish twigs.

The Cape region is also the habitat of a large number of ornamental plants such as some species of the genus *Mesembryanthemum*, the genera *Amaryllis*, *Clivia* and *Freesia*. The popular indoor plant African hemp *(Sparmannia africana)* also comes from this region (Ill. 37). Several odd orchids grow in the more open vegetational units in the Cape region. *Disa uniflora* (Ill. 41) is called the "pride of Table Mountain." It is the most beautiful species exhibiting the largest flowers of this genus of terrestrial orchids; it excels in a large number of different forms and grows on wet rock and swampy soils.

Blue-flowered orchids are very rare. Two of them, the terrestrial orchids *Herschelia purpurascens* (Ill. 39) and *Herschelia graminifolia* (Ill. 42), which are particularly striking and rare specimens, are found in the Cape region.

Primeval Madagascar

Though the island of Madagascar is separated from the east coast of the African mainland only by the roughly 350-km-wide Mozambique Channel its flora and fauna have little in common with the continent, because the island was separated from the continent in earlier geologic periods. Due to this isolation, many botanic curiosities have persisted on Madagascar. Thus, the island is rightly called a museum of living fossils. Many of its native plants have neither ancestors nor relatives in other parts of the world.

The well-known botanist Werner Rauh gave an impressive description of the flora of Madagascar:

A grand, horrible, and thrilling but at the same time pitiable spectacle is presented to the traveler who undertakes a journey through the island during the period from the end of October to the beginning of November. Large areas are in flames; from the mountains fires shine forth, and the flames greedily eat their way into the vegetation parched by a long drought, producing a howling and crackling noise.

One has to undertake troublesome and long journeys through almost inaccessible and rainy areas near the east coast and through dry regions of a semidesert character in the southern and southwestern parts of the island that covers a total area of 600,000 square km in order to experience the adventure of meeting with the last remains of a vegetation that is unique in the world.

In front of us there is an inaccessible thornbush. Almost every plant bristles with thorns such as the big spurges, thick-footed plants that form stems, namely, the Pachypodium species, a group of plants typical of Madagascar. Their barrel-shaped water-storing stems, three to five m in height, are densely covered with hard, prickly, stipular thorns. Predominant, however, are those eye-catching and queer thorny plants to which Madagascar owes its repute as an "aloof sanctuary." These are species of the Didieraceae family, trees or shrubs which remind us of the columnar cacti of South America because of their particular shape and outgrowth of strong thorns. A few representatives form actual forests whose survival, however, is imperiled.

It was only in the middle of our century that detailed information about the Didieraceae family was reported; their relationship was dubious for a long time, and it still remains a point at issue. It is a family of periodically leaf-shedding plants, with small inconspicuous flowers, that grow in the driest areas of Madagascar. Trees of the species *Alluaudia procera* (Ill. 43) reach a height of 10 to 15 m and form relatively large forests in the southwestern part of the island. These forests are imperiled by frequent bush fires, as pointed out by Rauh in the quotation from his report.

The plant that leaves the strongest impression certainly is the traveler's tree, *Ravenala madagascariensis* (Ill. 44). It is a plant of the banana family (Musaceae) up to 30 m in height, with a ligneous stem and large leaves in a fan-like arrangement. The stem reaches a height of only 3 to 6 m. During the rains, water is collected in large leaf sheaths, and travelers can use it as a reservoir during long dry spells. Frequently the traveler's trees form groves that survive the fires that rage every year. Because of its resistance, *Ravenala madagascariensis*, the only species of the genus, spreads over large areas in secondary forests. In many tropical countries it is cultivated as an ornamental tree. The large and conspicuous flowers are pollinated by small birds.

In the dry forests of western Madagascar grow several species of baobabs that have a somewhat antediluvian appearance. The giants among them on that island are *Adansonia grandidieri*, with trunks 30 to 40 m high and up to 6 m thick—trunks of monstrous growth (Ill. 45). The relatively small branch tops are quite disproportionate to the thick trunks, which, above all, serve for storing water.

The "star of the steppe," *Pachypodium lameri*, is a plant of the dogbane family (Apocynaceae), of the genus *Pachypodium* from the south of Madagascar (Ill. 46). All of the species of this genus are succulent shrubs. They grow up to 5 m high; they have a crown that does not branch out much on top and develops clusters of long leaves at the ends of the shoots.

Sclerophyllous vegetation in Australia

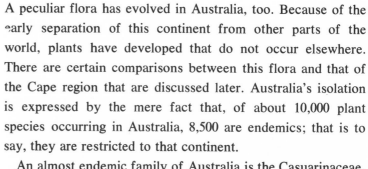

A peculiar flora has evolved in Australia, too. Because of the early separation of this continent from other parts of the world, plants have developed that do not occur elsewhere. There are certain comparisons between this flora and that of the Cape region that are discussed later. Australia's isolation is expressed by the mere fact that, of about 10,000 plant species occurring in Australia, 8,500 are endemics; that is to say, they are restricted to that continent.

An almost endemic family of Australia is the Casuarinaceae. which is composed of forty to fifty species, all of which belong to the genus *Casuarina* (Ill. 47). Sometimes this genus is called "kangaroo tree." A few species also occur in New Caledonia and Indonesia, and the area where only one species, *Casuarina equisetifolia*, is distributed ranges from Australia to India and Madagascar.

The habitats of this genus, which include trees up to 15 m in height as well as low-growing shrubs, usually are sandy soils or other dry soils of a poor nutritive value. Jointed stems and rodlike branches with whorls of scale-like leaves give them the appearance of giant shavegrass. Their flowers are small and of a simple form. Morphologically the *Casuarina* species of Australia are well adapted to the living conditions of arid regions. By the way, it is striking that only a few succulents occur in the arid regions of Australia.

The most widely known and most typical plants of Australia are *Eucalyptus* species of the myrtle family (Myrtaceae). These plants show many biological and ecological peculiarities, which are indicated especially by the structure of the flowers. The flowers have neither petals nor sepals, but the four coronal leaves grow into a cap, which is shed during the flowering time. The lower part of the flower bud forms a funnel-shaped calyx. The leaves are evergreen, leathery, and provided with oil pockets. Depending on the species, eucalyptus trees are of different sizes. Some of them are only a few meters high; others reach more than 100 m and, thus, are among the tallest plants of the earth. Their rapid growth is unparalleled in the entire plant kingdom. Eucalyptus trees are of particular importance in the drying up of large swamps infected by malaria, since their roots take in large amounts of water.

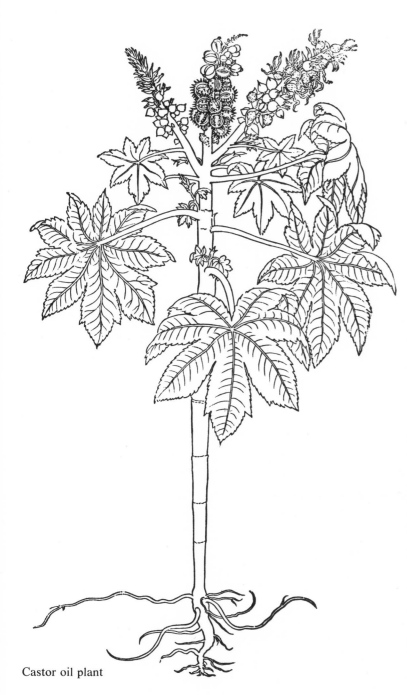

Castor oil plant

Karri forests consist mainly of the gum tree *Eucalyptus diversicolor*, which has a slender and straight stem 60 to 80 m in height (Ill. 50). During the drought, the flower buds which require two and one half years for their development, open. From pollination to maturity of the fruit a further one and one half years elapse. The seedlings, however, develop rather rapidly. During the first year, they grow 80 cm high; in the second year 3 m. During the following years, the annual rate of growth is 1 to 1.5 m. From about the twenty-fifth year, the rate of growth diminishes, dropping to roughly 30 cm per annum. On an average, the trees grow to be 350 years old.

On sandy heaths, which cover large areas in Australia, there occur several species of the family Proteaceae, to which we have already referred in connection with the Cape region. From this distribution of the family, many scientists derive a historical relationship between the floras of Australia and the Cape region.

Conspicuous representatives of Proteaceae are *Banksia coccinea* (Ill. 48) and *Telopea speciosissima* (Ill. 49), which have gorgeous flowers. A particular feature of *Banksia* is the mechanism of motion of the female flower parts. The styles form an arc when they emerge from the bud and, when the flower opens, they stretch, distributing pollen from the stamens. The woody fruiting inflorescences will open only after a fire. By means of a hook, the seeds remain attached to the wall of the fruit for several hours. Then they drop onto the burnt ground, which in the meanwhile has cooled, and there they find favorable conditions for germinating.

Grass trees of the genus *Xanthorrhoea* likewise show impressive forms; they are almost exclusively restricted to Australia and Tasmania. This genus belongs to the monocotyledonous plants and is closely related to the lily family. *Xanthorrhoea hastilis* has a relatively short stem crowned by a crowded tuft of 1-m-long, slender, rigid leaves that are arranged in nearly the same way as roof tiles and break off easily. The flower stalk, up to 2 m long, rises from the top of the stem and carries the spadix-shaped inflorescence (up to 50 cm long) with a large number of small flowers.

Green dragon

As has already been described, among the plants growing in deserts and semideserts, there are a few species that are adapted to extremely hard conditions of life, conditions under which most other species cannot survive. There, the limiting factors are dryness and heat. In this chapter we consider a few vegetational units with their typical plants where low temperature is the main limiting factor.

In the subpolar region of the northern hemisphere we find the tundra with a treeless vegetation and with short, cool summers. For the most part, the temperature is insufficient for trees to thrive, so that wide areas remain treeless. The rough climate makes particular demands on plants. They must utilize sunlight in the most efficient way. Stunted growth and evergreen leaves are typical features, and the rate of growth is small. Thus, the annual increase in length of the reindeer moss (*Cladonia rangifera*) is 1 to 3 mm. Twenty-year-old Siberian larches (*Larix sibirica*), among the few trees that grow in this region, hardly project from the low-growing herbs surrounding them. Their rate of growth is 1 to 2 cm per year.

The tundra may be divided into several different vegetational units. The forest tundra forms the transition from the continuous forest belt to the tundra vegetation proper. Predominant trees are birches (*Betula pubescens*), spruces and common pines. The dwarf shrub tundra, which is mainly covered with heather plants, comes next. Of all tundra regions, the Arctic tundra is situated nearest to the North Pole and supports a dense growth of mosses, lichens, and cushion plants. The density of the vegetational cover, however, gradually decreases northward until snow and ice prevent any plant life.

One of the most cold-resistant plants is the red saxifrage (*Saxifraga oppositifolia*), whose area of distribution extends to a point 750 km from the North Pole (Ill. 51).

Most plants of the tundra germinate in frost. The seeds must have been exposed to a temperature below the freezing point in order to be capable of germinating. The period of vegetation starts in June and ends in September. Because of the short period of vegetation, flower buds usually start growing in the preceding year, hibernate in a fully developed condition for eight to ten months, and burst into blossom immediately after

thawing. In this connection, the following phenomenon is of particular interest. Snow immediately surrounding plants frequently thaws earlier than snow located elsewhere, because a plant subject to isolation that penetrates the snow is heated more intensely. In this way, a cavity is produced round the plant that is closed on top by a thin transparent cover of ice, forming a tiny hothouse. In this cavity, the temperature may be up to 14 °C higher than on the surface of the ice cover, so that growth of the plants inside is favored extraordinarily.

A typical plant of the rock tundra is white dryas (*Dryas octopetala*) (Ill. 52), which, with its short, many-branched creepers, forms pads of several square meters in area. On steep slopes and on rocky debris, it stabilizes the soil remarkably well.

Extreme climatic conditions are also found in the taiga in eastern Siberia, where hot summers and very cold, almost snowless winters alternate. Temperatures as low as −60 °C have been recorded in the taiga. There grows, within the range of the continental pole of cold, the highly frost resistant Siberian larch (*Larix sibirica*), which sheds its needles and passes the winter in a resting state lasting several months. Its roots do not penetrate deep into the soil but are spread flat toward all sides because, in summer, the ground thaws to a depth of, at most, 50 cm.

Similar conditions and a similar zonation are found in the Alpine regions of the northern temperate zone where, with increasing altitudes, annual heat decreases, wintertime becomes longer, larger amounts of precipitation fall, and stronger winds blow. At the timberline, which is at an altitude of 2,200 m in the Central Alps and of 2,500 m in the Pamirs, the Altai Range, and the Tien Shan, a zone of competition has developed, where tree species of forests occur both isolated and mixed with other arborescent plants. At this line in the Central Alps we find the Alpine stone pine (*Pinus cembra*), with bizarre tops marked by storms, an interesting type of weather-beaten tree (Ill. 54). At altitudes of 1,600 to 2,200 m in Tien Shan, evergreen trees known as Tien Shan spruces (*Picea schrenkiana*) and some species of birch grow sporadically at the upper border of the coniferous wood zone (Ill. 55). The slender spruces with their narrow heads resemble dark green columns towering high into the blue sky, whereas birches display their autumn coloration by the end of August.

Where, trees are no longer capable of growing upright because of the severe climate, the region of crooked wood starts, above the timberline, with the mountain pine or dwarf pine *(Pinus mugo)* as the characteristic tree. The mugo pine, as it is also known, can do with a shorter vegetation period and is resistant to avalanches because of its shrubby spreading growth.

In the Alps, we find at this altitude the first Alpine roses *(Rhododendron ferrugineum)* with bright red flowers (Ill. 53). They lead over to Alpine meadows which are a charming sight because of their showy flowers. Alpine rose flowers are large, a striking feature of the plant, especially because they form a contrast to the vegetative organs of the plant, which remain low and grow in a matted manner. The plants nestle against the ground to escape from strong wind. At the same time, they must adjust to a more intense insolation, larger temperature differences between sunny and shady sides, and a very thick cover of snow on top of them in winter. Of ligneous plants, only a few species occur, including the dwarf juniper *(Juniperus sibirica)* and creeping species of willows (e.g. *Salix reticulata, Salix glabra*). Of herbs, species of saxifrage, including the red one, and species of gentian, are worthy of note.

Diphasium alpinum is a plant of high mountainous and Alpine regions of the northern hemisphere (Ill. 57). Outside of woods, this distinctly lime-evading plant is found on mat-grass turfs and mountain heaths. High mountains are marked by a peculiar vegetation not only in temperate latitudes but also in the tropics. The East African 6,000-m-high Mount Kilimanjaro occupies a special position among tropical mountains. The colossal massif towers high above vast African steppes. Climatic conditions are marked by distinct differences between day and night, whereas seasonal variations remain insignificant throughout the year. This has been conducive to the formation of curious plants. The most bizarre forms are found at altitudes from 3,500 to 4,000 m between low-growing cushion plants and the crooked-wood zone. At the Alpine level—i.e. at an elevation of about 4,000 m—one of the most impressive plants of tropical

high mountains, the lobelia *Lobelia deckenii*, grows (Ill. 56). A scape growing up to 2 m high, with the flowers arranged spirally about the stalk, arises from a graceful rosette of leaves. The flowers are cobalt blue, embedded in fleshy bracts, a favorite food of honey birds. Interesting flowers also occur in the South American Andes. The well-known cacti researcher Curt Backeberg reports on a trip through this region:

I go by train to Oroya, a mining place at an altitude of almost 4,000 meters. Near Ticlio, at an altitude of almost 5,000 meters, a thunder-storm raging over the peak announces the imminent drop of temperature, lighting and flashes and thunder performs a pandemonium in a gorge. Laboriously the train climbs the last gradient. Passangers from the coast desperately suck at an oxygen hose because sarroche or puna, the mountain sickness, here starts telling on them.

And then a snowstorm sets in. One can hardly see the next wagon. When the white flurries become transparent, now and then, ghostlike silhouettes of llamas, of those queer animals which in their dense coats feel cosy in the presence of storm and cold, sweep by. I intend to gather at the top of this range but, attacked by the fury of the elements, I am chased away. I go over to the warmer Rio Marañon; meanwhile the vigorous snowstorm will fade away.

After a few days I return. The weather has not yet cleared up. Pale clouds brood over the summit. Turbid pools mirror surrounding things in a dull green color, and it is no longer white everywhere. Here and there patches of snow seem to have withstood the rain following the snowfall.

And then I stand in front of the supposed remains of snow; they are cacti, *Thephrocactus floccosus*, the "living snow" as their discoverer rightly called the white furred plants. Stalwartly they spread in gullies exhibiting their coats as shaggy as those of the llamas which I saw in the snow-storm on the way here. What an efficient protection against stormy weather, many a man would think if he saw the white colonies. But this would be a fallacy; beside there is an accumulation of many-headed umbos with completely bare, dark-green tops, *Thephrocactus atroviridis*.

26 The strawberry tree *(Arbutus unedo)* is a characteristic plant of the maquis in the Mediterranian region.
27 A single flower of *Ophrys speculum* of the Mediterranean region

28 *Ophrys scolopax* also grows in the Mediterranean region.
29 At first glance one could not tell that *Serapias neglecta* is an orchid.
30 Inflorescence of *Serapias cordigera*

31 Inflorescence of the dwarf palm *(Chamaerops humilis)*, the only European species of palm

32 Red viper's bugloss *(Echium wildpretii)* is among the rarest of all plants.

33 The dragon tree *(Dracaena draco)* is the landmark of the Canary Islands.

34 *Euphorbia canariensis*, a type of spurge, grows densely in the Canary Islands.

35 The silver tree *(Leucadendron argenteum)* occupies only
a small area on the wet slopes of Table Mountain.
36 Buds of the snowball protea *(Protea cryophila),*
which occurs only on a mountain near Cape Town.

37 African hemp *(Sparmannia africana),* originating from
South Africa, is a popular indoor plant.

38 In the Cape region the large flowers (up to 25 cm) of *Protea barbigera* are reminiscent of fully mature pineapples.
39 *Herschelia purpurascens* is one of the few orchids with blue flowers.

40 *Protea cynaroides*, with its artichoke-like flowers, also grows in the Cape region.

41 *Disa uniflora*, the pride of Table Mountain, grows in the Cape region on wet rock and marshy soil.
42 *Herschelia graminifolia*, with its blue flowers, also originates in the Cape region.

43 The species *Alluaudia procera* of the Didieraceae family covers large areas in southwestern Madagascar.
44 The traveler's tree *(Ravenala madagascariensis)*

45 The baobab *(Adansonia grandidieri)* grows in the wild woods of western Madagascar. The specimen depicted is the biggest in the world.

46 The stems of *Pachypodium lameri* grow up to 5 m high.

47 Plants of the genus *Casuarina* (coextensive with the family Casuarinaceae) are characteristic of Australia.

48 *Banksia coccinea* is found in the Australian heaths.
49 *Telopea speciosissima* also is typical of Australian flora.

50 Eucalyptus trees occur only in Australia. The trees in the illustration are *Eucalyptus diversicolor*.

51 Red saxifrage *(Saxifraga oppositiofolia)* is the northernmost of all phanerogams.
52 White dryas *(Dryas octopetala)* is typical of the rocky tundra.

53 Like many other Alpine plants, the Alpine rose *(Rhododendron ferrugineum)* attracts attention because of its shining colors.

54 The stone pine *(Pinus cembra)* and the Alpine rose *(Rhododendron ferrugineum)* occur at the forest boundary in the central Alps.

55 Spruces *(Picea schrenkiana)* and birches at the forest boundary in the Tien Shan Montains of Central Asia.

56 *Lobelia deckenii* grows on Mount Kilimanjaro at an altitude of about 4,000 m.

On the lookout for pollinators

Trapped pollinators

Useful beauty

Diversity of orchids

Fooled males

Complicated forms of fertilization

Birds and bats

Opening times of various flowers

For a long time man has paid particular attention to the great variety of flower shapes and colors. Only after the discovery of sexuality of plants at the end of the 17th century, however, was the secret of structure and pollination of flowers unveiled. Ome of the pioneers in this field was the Berlin botanist Christian Konrad Sprengel (1750–1816), who thoroughly studied flowers and published the results of his painstaking work, conducted for a period of many years, in his book *Das entdeckte Geheimnis im Bau und in der Befruchtung der Blumen* (1773). Although Sprengel's thinking was strongly influenced by teleology, his studies and their results are of great value even today.

With the triumphant advance of anthropogenesis, a new epoch of biological researches of flowers began. No less a person than Charles Darwin (1809–1882) made considerable contributions to these researches, especially by his works on the pollination mechanisms of orchids (1862). As a result of the classical studies on the interrelations between the honeybees and flowers by Karl von Frisch as well as the comprehensive analyses of the individual visitors of flowers by Fritz Knoll, experimental flower ecology was developed during the first half of the 20th century, which considerably enlarged man's knowledge of pollination mechanisms.

The flowers of angiosperms consist of short sprouts of a limited growth. Commonly a flower is composed of a usually green calyx, a mostly colored corolla, stamens and carpels. Sepals and petals together form the perianth, the external envelope of a flower, which protects stamens and carpels. This was the basic structure from which a host of flowers of a great diversity were derived by changes in the perianth, and in the number, shape, size, and color of petals and sepals, as well as by changes in the position of the individual flower parts, by connations and adnations—reduction or increase in number of flower parts—and by the formation of odorous substances and nectaries.

All of these changes, in the last analysis, serve for sexual reproduction of plants, which is dependent upon pollination of the stigma and fertilization of the ovules. During fertilization, the male germ cell contained in pollen, with which it was placed on the stigma, unites with the female germ cell, the egg, enclosed in the ovule. Pollination is differentiated into various forms. We speak of self-pollination when pollen is transferred from the anther of a flower to the stigma of the same flower, whereas pollination of one flower by another growing on the same plant is called geitonogamy. Usually cross-pollination takes place; in this case, the pollen transferred to the stigma of a flower comes from another plant.

For pollination, plants are dependent upon certain carriers, namely, wind, water and various animals. Below we shall deal only with zooidogamous plants, because they show the largest range of different adaptation phenomena. Insects are the main pollinators. In the tropics and subtropics, more than 2,000 species of birds, especially humming birds, nectar birds and honey eaters, as well as several bats play a role in pollination.

Striking forms, colors and odors of the flowers attract the visitors. The animals search for food on the flowers, especially nectar, pollen, and hair. Other reasons for frequenting plants are the gathering of odorous substances, material for nest building, deposition of eggs, and sometimes mating behavior. Although the interrelation between flower and pollinator is of a temporary nature, it is vital to both of them. They derive benefit from each other for survival of the species. The importance of this interrelation is demonstrated by the fact that, if there were one year without flowers, more than 1,000 species of insects and of some birds (hummingbirds, honeysuckers, and honey eaters) would die out, because these animals feed on flowers only and are completely dependent on them. For attracting animals, plants use different methods. Let us take a closer look at a few special flowers and their methods of attraction.

Edithcolea grandis (Ill. 58), producing large flowers that reach a diameter of up to 15 cm, is frequently called "queen of the succulents." The area of distribution of this beautiful succulent is steppe-like bushland in Tanzania, Kenya, and Somalia.

As to pattern and color, the flowers resemble a Persian carpet. On a pale yellow ground, an artistic pattern of purple-brown color can be seen. The flower has the shape of a shallow plate, with five wide lobes that terminate in tips bent back. Five rays of movable claviform hairs run from the center of the flower to the five indentations in the edge. The dark margin of the flower is also provided with small purple hairs. The plant creeps between brown-gray boulders and red-brown sand, frequently much branched toward all sides and hanging down from rocks; during dry spells it takes on the same color as the environment. To protect the plants from animals feeding on them, many gibbosities with sharply pointed thorns grow on their hard, five-edged, hairless stems.

The South African succulent *Decabelone grandiflora* (Ill. 59) attracts pollinators by movable glimmering bodies. The large funnel-shaped bells of the flower are light-yellow in color and are covered with brown-red spots and furrows. The corolla appendage ends in ten filamentary lacinias, and a spherical flesh-colored button is suspended at the tip of each of them. Very slight vibrations or movements of the air cause the buttons to swing like a pendulum; this produces an amazing glimmering effect.

The tuber-forming *Brachystelma barberiae* (Ill. 64) is known as the "window flower plant" because of its particular floral structure. The flowers are arranged in an umbel, and the lacinias of the wide, bowl-shaped corolla are extended and their tips remain connected with each other. In this way, five window-like openings are brought about. The flowers are black-violet, showing a lighter shade at the base.

Remarkable inflorescences are found in plants of the genus *Dorstenia* (Ill. 65), which is native to tropical regions. Like all other plants of the mulberry family, this genus has unisexual flowers that are arranged in bizarre inflorescences. The flowers are small and inconspicuous, forming a dense cluster on a chocolate-brown cushion shaped like a flattened shield, which is surrounded by thin bristle-like bracts arranged in a star. During ripening, cavernous tissue is formed in the flower cushion, which exerts pressure on the floral leaves to the effect that the fruits, small nuts, are flung off a few meters.

Tropical beauties of particular charm are the showy flowers of the genus *Costus*, plants of the ginger family. One stamen of the flower is turned into a display organ—a large, colored labellum, having the shape of an upside-down egg—while the filaments of the stamens are broadened like petals. The species *Costus speciosus* (Ill. 60) develops an upright cone-shaped inflorescence that bears a large yellow flower on top. The edges are folded back toward the outside and show striking red-brown stripes.

Some flowers "hide" the nectar at the flower bottom so that insects have to perform pollination while they search for nectar. Such devices are found in passion flowers (genus *Passiflora*); they are tendril-bearing vines comprising a large number of species that occur especially in the tropical virgin forests in America, Asia, Australia, and Oceania. The flowers have a queer but appropriate structure. Between corolla and five stamens, arranged on a column and three stigmas, there is a filamentary circular member called a fringed corona that closes the cup-shaped bottom with the nectaries of the flower. Nectar is accessible only through a narrow annular gap that cannot be reached by small insects. Pollinators must possess a long proboscis to get down to the nectar. The relatively big creatures eagerly move on the fringed corona, touching the sexual organs, which are at a higher level, with their backs. In this way, pollination is effected.

The fringed corona is different in shape and color with the various species. In granadillas *(Passiflora quadrangularis)*, the petals are reddish and the long threads of the fringed corona are white and purple (Ill. 63). In *Passiflora racemosa* (Ill. 62) the white fringed corona effectively contrasts with the scarlet corolla.

Blossoms of carrion flowers (plants of the genus *Stapelia*), included in the milkweed family, have no regard for their visitors. Most of the flowers emit such a strong smell of carrion that carrion-flies like to deposit their eggs or larvae inside them; but eggs and larvae perish, because they will not find any food. As the plants are widely scattered in the desert regions of South-West and South Africa, the pollinators will not be eradicated. *Stapelia flavopurpurea* (Ill. 61) puts forth flowers that have a diameter of only 3 cm. They are flat, shaped like a wheel, and quinquefid. Their color is an eye-catching dark yellow interrupted by irregularly arranged cross calli. The center of the flower is white and covered with red claviform hairs. It should be noted that this flower, which is native to the Karroo in South Africa, does not smell of carrion—as all of the other species of this genus do; instead it emits a pleasant smell of honey.

As the above description has alredy shown, the enticement of insects depends not only on color but also in a high degree on smell. The smells involved are not in every case the pleasant kinds produced by violets and roses; there are also evil-smelling species of plants, above all the aforesaid carrion flowers. These belong to several families, especially the arum family (Araceae), the birthwort family (Aristolochiaceae), and the milkweed family (Asclepiadaceae). Carrion flowers entice flies and beetles with a yellowish-brown or black-red color and a smell that is offensive to us, sometimes nauseating. In this manner they feign the presence of rotten meat on which these animals usually feed or where they deposit their eggs. The extraordinary way in which some plant species secure pollination will be shown by a few examples.

Inflorescences of plants of the arum family are marked especially by the size and color of the frequently tubular sheathing bract or spathe that encloses the base of the spadix, the typical form of inflorescences of this family. To illustrate the function of the so-called kettle trap flower, an example of the cuckoopint *(Arum maculatum)*, growing in the shady deciduous forests of Europe, is given here (Ill. 66, 68). In the period from April to June, the plant develops a purple spadix which is enclosed in a large greenish-white spathe. The lower part of the latter forms a "kettle." Its entrance is closed by bristles of the spadix pointing downward. Under them the male flowers and further down the female ones are arranged. Due to chemical processes, the temperature in the lower part of the inflorescence becomes higher than in the other part; this heat, together with odorous substances, attracts insects for pollination. Small flies, especially the early swarming owls, creep into the trap. Hindered by the rigid bristles, they cannot get out of it. Up to 4,000 of such small gnats have been found in one trap. Usually, the animals arrive with a load of pollen from a plant visited before. When creeping about in the kettle, they pollinate the female flowers on the lower part of the spadix. After pollination, the rigid bristles become slack, so that the insects are allowed to leave the more or less hospitable kettle trap. As the insects depart, the stamens open and strew the animals with pollens which will be carried to other flowers.

The tropical and subtropical representatives of this family possess inflorescences that, because of the size and shape of the spathes and processes and appendages of the spadix, have a very queer appearance.

The genus *Amorphophallus*, composed of about 80 species, is among such queer fellows of the plant kingdom. They are native to the tropical forests on the Indonesian islands, in India, and in Sri Lanka (Ceylon). The inflorescence of the aroid known as krubi *(Amorphophallus titanum)*, discovered in Sumatra in 1878, sprouts rather quickly from a tuber that attains a diameter of 50 cm and a weight of 34 kg, because an ample supply of nutritive substances is contained in the tuberous reserve. In a few days it can attain the imposing height of 2 m. The spathe, up to 1.3 m in size, is brown-purple in color with a greenish shade inside and has a tubular base. The smell of carrion emanates from the flowers, attracting hosts of flies. In Sumatra explorers observed that even elephants act as pollinators, for as they drink from the flower, they rub pollen from the staminate flowers with their foreheads and later transfer it to another plant. After pollination the aerial parts of the plant rot and fall off. Thereupon the single leaf of the plant appears; it has a petiole that attains a length of 5 m and a thickness of 10 cm. The blade, whose sections may become 3 m long, unfolds like an umbrella.

Amorphophallus eichleri (Ill. 67), native to tropical West Africa, is of a similar structure but not so big.

The bizarre flowers of birthwort, genus *Aristolochia*, also form traps resembling a kettle or the like; the 300 species of this genus occur primarily in tropical and subtropical regions of America, Africa and Southeast Asia. The united petals of the flower end in a tube, which is enlarged to form the floral kettle. A striking feature of the perianth is a colored border that is used as a place of rest by approaching insects. The smell of carrion emanates from the narrow opening of the flower. In 1928, Erwin Lindner described the complicated structure of the flower of *Aristolochia lindneri* as follows:

The most essential part of the flower is found behind the large display apparatus, the lower lip. In the place of a plain tube with a pilose inner surface found in other species, this one has an angular vestibule with almost smooth walls leading up to a vertical diaphragm closing the space behind and providing only a funnel-shaped opening through which visitors can enter to get to the bottom of the flower into the space above the torus or receptacle. Except for this opening, this space, the kettle trap, is fully enclosed; a smell so irresistible to carrion flies emanates from it; and it houses the column of sexual organs on its bottom. To show the guests the way through the dark and angular vestibule, the dark brown color is left out in those places through which they have to pass in order to arrive at the small opening in the slightly spotted diaphragm through which the light shines from the space behind. Its walls are but slightly spotted, and the relatively intense light feigns a back door, a loophole for the hesitating intruders to escape into the open.

The sexual organs are at the bottom of the kettle. They are surrounded by a translucent annular window whose effect is intensified by a dark annular margin. When insects are approaching this window, mature stigmas strip pollen from their bodies.

Aristolochia grandiflora, the pelican flower, develops flowers that often attain a diameter of 30 cm and more, being among the largest flowers of the plant kingdom. Their lilac, slightly iridescent ground color is interrupted by red-brown spots, and the black-brown throat is bordered by a pale ochre yellow color. The tailed fringe, hanging downward, often attains a length of 60 cm. *Aristolochia brasiliensis* (Ill. 71) is quite different in color and shape; it exhibits a surprisingly elegant lineation and a fine net-like nervation. The flower entrance of *Aristolochia elegans* (Ill. 70) is conspicuously marked by contrasting colors. Its center is yellow and its margin purple.

The flowers of plants of the genus *Ceropegia* of the milkweed family have quite peculiar structures. In them we find an ingenious combination of the kettle trap and other ecological floral devices. Besides the smell of carrion, interesting, usually antenniform, glimmering bodies serve for attracting. Also in addition to the kettle trap, a clamping trap is arranged in the flower. These flowers are veritable works of art produced by nature; there is no other genus of plants that develops such bizarre forms. From the small flower of *Ceropegia cancellata*, having a size of only 1.5 cm, to the "parachute" of *Ceropegia sandersonii*, which attains a length of 8 cm, the most remarkable flower shapes and color variations are found in this genus composed of more than 150 species native to Africa, Madagascar, India, and Southeast Asia. Two particularly conspicuous species are described below. *Ceropegia galeata* (Ill. 72) has relatively large flowers, with kettle-shaped tubes at their bases. This tube narrows in the central portion and flares upward, forming an umbrella fringed with purple cilia that produce a glimmering effect. Their colors vary from gray-green with brown-violet dots to yellowish shades.

The flowers of *Ceropegia distincta* var. *haygarthii* (Ill. 73) are mottled with light gray-green and ruby red colors. Their funnel-shaped opening is formed by five windows. A small ciliate "lantern" carried by a pedicule is arranged above them.

In all *Ceropegia* species hairs growing inside the tube and pointing downward serve as traps hindering insects from creeping out. After pollination, however, the perianth wilts, releasing the animals.

A combination of sliding, light, and clamping traps is presented by the red-brown lady's slipper *(Cypripedium calceolus)*, an orchid that occurs in Central Europe and is protected in many countries (Ill. 69). The flowers, borne on slender stalks, rise singly, or rarely in pairs, above large oval leaves. The golden-yellow lip, which has the shape of a slipper, contrasts impressively with the surrounding petals. The spectacular effect of the yellow lip is increased by transparent windows. The edge of the lip is curved inward and, like the entire slipper, is smooth and bright, so that approaching insects do not find a hold and, consequently, slide into the slipper.

When creeping out of it, the insects, compelled to climb over sort of a "picket fence" of small hairs, are pressed against the stigma, stripping off pollen brought along with them on their backs from other flowers. Then the animals force through the narrow opening of the lip, at the same time picking up some of the sticky pollen.

Habitats of lady's slipper are open woods and shady deciduous forests, open pine or mixed forests, and clear scrubland. Usually they grow in calcareous soil. The general area of distribution ranges from Central and North Europe to the Caucasus Mountains and to Siberia. In the Mediterranean region, this species is not found.

In his report on his journey from the Ural Mountains to the Altai Range, undertaken in 1829, Alexander von Humboldt gives details of the hosts of lady's slippers growing in the Ural region, numbering hundreds of thousands of specimens. Today, lady's slipper has become rare in its entire area of distribution.

The above-mentioned plants trap insects for the purpose of pollination and usually release them after they have done their duty. Thus, these plants are clearly distinguished from insectivorous plants, which will be dealt with further below.

Insect orchis

Hardly any other family of plants can rival the orchids with diversity of lower form. The habits of orchids, their forms of adaptation and the specialization of their flowers are of special interest to the botanist, for they have resulted in species of such an extraordinary beauty that orchids occupy a special position among ornamental plants. Whereas European orchids, with the exception of lady's slipper, have relatively small flowers, many tropical representatives fascinate by the size of their flowers alone. Orchids are not only one of the most wonderful and interesting families of plants but also the richest in species, comprising some 25,000 on all continents. In view of this great abundance, it is not possible to give an exact survey of orchids in this book and we ask readers to be satisfied with a very small selection. For readers particularly interested in this subject we refer to the comprehensive special literature.

In the tropical rain forest on the eastern coast of Madagascar grows the orchid *Angraecum sesquipedale* (Ill. 74), which bears a particularly long lower spur. The perfectly white flowers, which attain a diameter of 15 cm, possess a spur 20 to 30 cm in length, the longest found in the plant kingdom. At the bottom of the spur is nectar, and butterflies, attracted by an intense smell of vanilla, suck it up with proboscises that can be extended to the desired length, and at the same time pollinate the flower. For a long time, this strange flower has been enigmatic to botanists and entomologists, because one did not know a butterfly having such a long proboscis. In 1862, Darwin received this plant from Madagascar and he predicted the existence of such a butterfly. The butterfly was finally discovered, though only in 1903, and it was given the name of *Xanthopan morgani praedicta*, meaning "the predicted."

One of the most curious orchids is the Central American and South American genus *Catasetum* (Ill. 76). Darwin called this plant the most remarkable of all orchids. The flowers are unisexual and largely vary in form. They have an ingenious device for pollination. Closer examination of a male flower shows that in its central area it has two thread-like processes that point downward, approaching the concave fringed labellum. These antennae are connected with the pollinia and, when touched by an insect, the mass of pollen is catapulted out and clings to the visiting insect by adhesive disks. During a subsequent visit to a female flower, the pollen mass is easily stripped off from the insect by a transversely arranged sexual colum within the female plant.

Another particular feature of these flowers is the fact that they do not supply the Euglossane male (genus *Euglossa*) with nectar, as is usually the case, but with a scent secretion produced by the gland tissue of the labellum. The male insects collect this liquid and store it in special organs. According to investigations conducted by Stefan Vogel (1966), this flower scent is used by the males for marking their flight territory where they intend to court.

The flowers of the genus *Stanhopea* (Ill. 75) of epiphytic orchids indigenous to Central America, Peru, and Colombia are strikingly, almost grotesquely, shaped. The fact that these flowers grow downward from the cluster of roots is a feature queer enough, but what are really fascinating are the unusual structures of the flowers, their colors, and their overpowering scent. The labellum, or lip, is a very complicated display apparatus and points downward like the sexual column, whereas the five other sepals and petals are turned back. The two oddly rounded, somewhat concave sepals stand out, while the central one grows upward. The two lateral petals are turned back as far as possible. All petals are light yellow in color with numerous dark purple spots. The thick fleshy lip is made up of three sections. Of these, the rear one is opened out like a sac, forming an open-ended cavity toward the sexual column, the inner glandular tissue of which is used by insects for feeding. As they crawl out of this cavity, insects lose their footing on the smooth surface of the lip and slip down as if on a slide, stripping the pollinia or brushing against the stigmas where they leave pollen brought along with them. The front part of the lip has two horn-shaped growths. That is why these plants are called *toritos* (little bulls) in their native lands.

When the bud of an orchid unfolds, an interesting turning or twisting of floral parts takes place. In its bud, the orchid flower is in an inverted position. When unfolding, the stalk of the flower and the ovary perform a twist of 180 degrees, the so-called resupination. Consequently, the individual parts of the flower are turned upside down, so that the lip, the member of greatest importance to the visiting insects, becomes the lower part (Ill. 80, 81). In the lady's slipper, the individual flower is turned upside down, and in many tropical epiphytic orchids, the torso is twisted (Ill. 82, 83).

In the small terrestrial orchids *Malaxis monophyllos* and *Hammarbya paludosa*, resupination is through 360 degrees so that, after turning, the flower is again in its original position.

In *Epipogium aphyllum* turning is not effected, so that the lip is directed upward during unfolding (Ill. 79). This leafless saprophyte (a plant living on humus) grows in shady and damp deciduous woods between decaying leaves and in spruce forests on moss and on a layer of decaying needles on top the soil. It is distributed from South and Central Europe to Siberia and to the Caucasus Mountains. It multiplies mainly vegetatively by runners from the rootstock. Flowering may be discontinued for periods from several years to a few decades, and then the plants suddenly appear from the brown soil.

Whereas in *Epipogium aphyllum*, non-turning of the flower stalk is typical, this phenomenon occasionally occurs in other species of orchids, which then appear to be abnormal. This can be observed, above all, in *Orchis purpurea* (Ill. 77, 78).

Unusual floral shapes are present in plants of the orchid genus *(Ophrys)*, whose 22 species are found mainly in the Mediterranean region but are also distributed in Central and western Europe, including Scandinavia, in western Asia and in North Africa. Shape and color of the lips differ largely from those of the other petals. They resemble an insect body, especially because they have a velvet-like appearance and show a distinct mark. This genus is characterized by a particularly interesting, unique method of pollination. In contrast to most of the other flowering plants, species of *Ophrys* do not entice pollinators by pollen and nectar, because they do not produce them, but exclusively by the form of, and the scent emitted by, their flowers. Actually they are highly specialized dummies of insect females—that is to say, they are taken for such by males of various insect species. From this it follows that these flowers do not excite the feeding instinct but rather the reproductive instinct serving the preservation of the species. As early as 1917, A. Pouyanne in North Africa and M. J. Godfrey in England found that only males of bee and wasp species visited the flowers of these orchids. At that time, an explanation of this fact could not be given. In the past 25 years, however, B. Kullenberg and other scientists have studied the interrelations between flowers and insects, using *Ophrys* species for this purpose.

Above all, males of digger-wasps (Sphecidae), ground-bees (genus *Andrena*), dagger wasps *(Scoliidae)*, and ichneumon flies of the genus *Eucera* are attracted by coloration and shapes of the lips as well as by the scent emitted by the flowers. Obviously, the smell plays a special role because it corresponds with the species-specific sexual scent of the female insects in question, causing the male to visit the flower. The shape of the lip entices the male to alight on the flower, and it tries to copulate. While it moves about in search of the female, pollination is effected. While in other plants, insects receive a reward in the form of nectar or pollen, this is not the case with this plant. The insects are fooled, the flowers offer nothing but an empty promise.

Of the orchid species present in Central Europe the fly orchid *(Ophrys insectifera)* is the most widely distributed (Ill. 84) because its area of distribution extends to Scandinavia. Its lip resembles a colored fly resting on a greenish flower with its antennae extended. As pollinators, only males of the digger-wasp *Gorytes mystaceus* have been observed so far.

The spider orchid *(Ophrys sphegodes)* is a Central European orchid which starts flowering on sunny calcareous slopes and in dry meadows from the end of April, thus being the *Ophrys* species that unfolds first (Ill. 85). Shape and pattern of the lip closely resemble a spider.

In the Mediterranean region, the bee orchid *(Ophrys apifera)* is pollinated by the male bees of the genus *Eucera* (Ill. 86). This plant is also adapted to self-pollination if suitable insects are missing. This is the case in Central Europe because this small orchid immigrated to this region only in the postglacial warm period. There self-pollination is effected by pollinia of one flower which drop on the stigma of the same flower. Fortunately the species is still found in relatively many places.

Ophrys holosericea, which flowers from the end of May to the end of June, possesses the largest flowers of the Central European species.

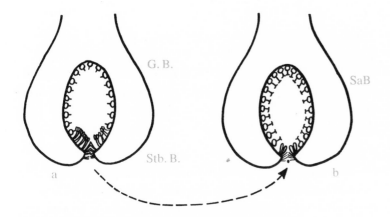

Schematic representation of fructification in figs:
a Inflorescence with male (St.B.) and gall flowers (G.B.)
b Inflorescence with female flowers (Sa.B.)

The interrelations between flower and insect are of a quite distinct nature with the fig tree *(Ficus carica)* (Ill. 87). The urn-shaped inflorescence axis bears small individual flowers on its inner surface that are accessible only through a narrow opening. The sexes are clearly separated. There are plants having female flowers only, whereas other plants develop male flowers in the upper half of the inflorescence and degenerate pistillate flowers, known as gallflowers, in the lower half. Small gall wasps of the species *Blastophaga grossorum* visit the inflorescences and deposit their eggs in the gallflowers. Their offspring leave the inflorescence loaded with pollen and fly to female inflorescences of other plants, pollinating the mature stigmas. As the stigmas are relatively long, the insects cannot introduce their ovipositors into the ovary of the female flower and therefore do not deposit eggs inside these flowers.

In ancient times Theophrastus and Pliny pointed out that, beside the fig trees yielding fruit, wild figs should be planted because small insects are inside the wild figs and move to cultivated figs, where they are instrumental in the fertilization of the flowers so that fruits can ripen. In Egypt it was a ritual to place flowering branches of the caprifig *(Ficus carica sylvestris)* in a garden of flowering fig trees to facilitate pollination, a horticultural operation which is known as caprification. The gall wasps, also called fig wasps, carry pollen from the flower of the caprifig to that of the edible fig. It is interesting that the gall wasp was found in figs from Egyptian tombs buried 3,000 years ago.

Probably, interrelations between plant and animal are strongest in *Yucca* species, plants belonging to the agaves, of which about 35 species are found in the arid areas of the southern U.S.A. and in northern Central America. The large, showy, white, bellshaped flowers of Adam's needle *(Yucca filamentosa)* form terminal panicles (Ill. 88). The flowers are pollinated by the small white moth *Pronuba yuccasella.* Toward the evening, the flowers open, emitting a strong smell. This attracts female moths which collect pollen in the flowers, forming it into a ball. When visiting another flower, they lay eggs into the ovary through their long ovipositor and then press a part of the pollen into the stigmas. The larvae hatching from the eggs eat some of the ovules. In the other loculi of the ovary, however, seeds develop that serve for multiplication and thus for perpetuating the species. Whereas part of the flower is eaten by larvae, the female moth ensures pollination and thus multiplication. In this case, flower and insect are highly specialized and largely adapted to each other.

Oleander

In tropical and subtropical regions of the earth, not only insects but also birds play an important part in the pollination of flowers. One reckons with about 2,000 of these species of birds, including hummingbirds in America, nectar birds in Africa, Southeast Asia and Australia, as well as honey eaters in Australia and New Guinea as the most important representatives. Like flower insects, the flower-visiting birds are also marked by peculiarities and the flowers visited by them show various forms of adaptation to their pollinators. First and foremost, the animals come to the flowers to feed on nectar and to look for water. In addition, they also eat pollen and small insects. These creatures are more eager and successful exploiters of flowers than the insects are. One has observed that such a small animal as a hummingbird undertook 42 flights for feeding within three and one-half hours, visiting roughly 1,300 flowers. Further, it has been calculated that a hummingbird must visit more than 1,000 flowers of fuchsia to cover its daily energy requirement.

As very skilled and fast flyers they hover in front of flowers, performing up to one hundred beats with their wings in a second, dip their long and pointed bills like probes into the calyces, rapidly extending and retracting their long, slender tongues, which are somewhat forked at the tip, and lick droplets of nectar in rapid succession while they remain to hang fluttering in the air. Bird flowers are distinguished from insect flowers by the usually more abundant supply of nectar and larger size. Birds are attracted by gay and striking colors. A particularly glaring red is preferred because it produces a strong influence on birds as a signaling color.

The large funnel-shaped flowers of the *Hibiscus* species, of which more than 250 grow in the tropics, are probably among the most beautiful bird flowers. The showy flowers of the species *Hibiscus regius* occur in Hawaii where the hibiscus is the state flower (Ill. 89). A large-flowered breed is "Flamengo," which has yellow-white flowers with a red throat. Flowers in red and pink shades are used as a hair ornament. Usually they grow separately on the branches, are relatively large and usually have five petals. A long style with stigmas, which also bears the anthers, rises from the center.

Another widely known species is the Chinese rose *(Hibiscus rosa-sinensis)*; a large number of beautiful varieties have been obtained by breeding (Ill. 92). Of particular charm are the "ear-drops of the princess" *(Hibiscus schizopetalus)*, which grow in tropical East Africa (Ill. 91). The long pendant flowers have petals that are bent back, deeply slotted, and of an orange-red to red color. Anthers and styles project from the flower for an unusually long distance. In their native country, nectar birds (especially *Cinnyris* species) visit the attractive flowers. Hovering in front of a blossom, the nectar bird introduces its long bill into the nectaries in the calyx. During this action, it first contacts the stigma and then the anthers. The relatively big pollen grains, which have tiny prickers, cling to bill and feathers, and are transported from one flower to another.

For the bird-of-paradise flower *(Strelitzia reginae)* of South Africa, plants of the Musaceae family resembling bananas, nectar birds also are the pollinators (Ill. 93). The particular features of this flower are shape, color composition, and pollination mechanism. Four to six flowers grow in a boat-shaped spathe, which is borne on a long stalk. One flower after another is put forth from the gray-green sheath and each unfolds in the course of several days. The color of the orange-red sepals and of the sky-blue petals is impressive. Two of the petals connate into a lanceolate tube in which the anthers are arranged and through which the stigma grows. When alighting, the honey birds press the floral tube open. Consequently, the anthers become accessible, and when the birds further penetrate into the flower, they pick up sticky pollen on their abdomen and strip it off on the projecting stigma of the next flower.

In the case of the coral tree *(Erythrina crista-galli)*, the flower typical of the papilinaceous plants is formed into a large and wide display organ with a dish-like impression (Ill. 90). Stamens and styles project slightly from it. The dark cherry-red color of the flower attracts nectar birds for pollination. By governmental decree, the coral tree was declared the national flower of Argentina. A similar decree also exists in Uruguay. The tree grows in South America in gallery woods on humid soil. Because of its conspicuous flowers, it is cultivated in parks and, especially in Europe, in tubs.

Among mammals, only bats are of greater importance as pollinators. Being night animals, they need not be attracted by conspicuous floral colors. Instead, the flowers emit a sourish-musty smell, excrete large amounts of mucous nectar, and produce large quantities of pollen. The plants are nocturnal flowers. The flowers must be located freely above the leaves or be suspended on long stalks to be available for bats.

A typical bat flower is *Marcgravia evenia* (Ill. 94). These liana-like tropical plants form hanging inflorescences at the end of the shoots. Striking features are the jug-like nectar leaves, with small flowers arranged above them.

Bananas (genus *Musa*) are also visited by bats. The flowers resemble large hanging clusters of spikes. In them, flowering takes place from the bottom to the top. Every evening, one bract unfolds so that two rows of the light-yellow, unisexual flowers become free. Attracted by the smell, which is disagreeable to men, the bats cling to the sturdy bract during feeding, effecting pollination.

Opening times of various flowers

The opening times of various flowers are also of particular interest. Although most flowers open in the morning and remain so throughout the day, there are a number of exceptions. The well-known Swedish naturalist Carl von Linné (1707–1778) was among the first who studied the opening and closing movements of plants. In 1770, he arranged a so-called flower clock (*Horologium florae*) in the Botanical Gardens of Uppsala, which showed some typical opening times of flowers. The following survey shows the opening times of a few selected day-blooming plants:

4 o'clock tall morning glory *(Ipomoea purpurea)*
5 o'clock corn poppy *(Papaver rhoeas)*
     pumpkin *(Cucurbita pepo)*
6 o'clock fire weed *(Epilobium angustifolium)*
     chicory *(Cichorium intybus)*
7 o'clock coltsfoot *(Tussilago farfara)*
8 o'clock marsh marigold *(Caltha palustris)*

In contrast to these daytime flowers, there are several plants that unfold their flowers in the evening or even during the night. Among these are:

6 o'clock evening primrose *(Oenothera biennis)*
7 o'clock catchfly *(Silene noctiflora)*

Plants that open their flowers at night are thornapple species (genus *Datura*) of the family of nightshades (Solanaceae). The large trumpet-shaped flowers open toward evening (7–8 o'clock P.M.) and remain open for about 24 hours.

It is interesting that the intense scent is emitted by these flowers only toward evening, and they are visited only by moths. These moths have a long proboscis enabling them to reach the nectar at the bottom of the ovary. Particularly striking flowers are shown by *Datura suaveolens* which grows in Brazil (Ill. 95). These trumpet-shaped flowers are white and grow up to a length of 30 cm.

From 9 P.M. to 3 A.M. representatives of the cacti genus *Selenicereus* open their large flowers. The actinomorphic flowers of the "queen of the night" (*Selenicereus grandiflora* and *Selenicereus macdonaldiae*), which have a strong smell of vanilla and reach a diameter of up to 40 cm, are very impressive (Ill. 96). The outer floral leaves are of a bluish orange-yellow color. These cacti, which are native to Jamaica, Cuba, and Haiti as well as to Argentina, are cultivated in display hothouses as well as by many cacti fans as an attraction throughout the world.

The flowering periods of the plants differ greatly. Some plants flower for one day only, sometimes only for a few hours. Among those one-day flowering plants are European wood sorrel (*Oxalis europaea*, 8 A.M. to 4 P.M.) and the heron's bill (*Erodium cicutarium*, 8 A.M. to 5 P.M.). In contrast, the flowering time of cranberry *(Vaccinium oxycoccos)* is 18 days. With other plants, the flowering period may even last for more than one hundred days. It is, for example, reported that the panicles with more than a hundred blossoms of the Malayan flower *Phalaenopsis amabilis*, an orchid growing on the Sunda Islands, have kept for four months (Ill. 97). The flowering period of other tropical and subtropical orchids also extends from several weeks to two months. Examples are *Oncidium cruentum*, with 60 days; *Paphiopedilum villosum*, with 70 days; and *Odontoglossum rossii*, with 80 days (Ill. 98).

58 With their yellow and purple-brown colors, the flowers of *Edithcolea grandis* form a gay carpet.
59 Yellow and red are the colors of the South African *Decabelone grandiflora*.

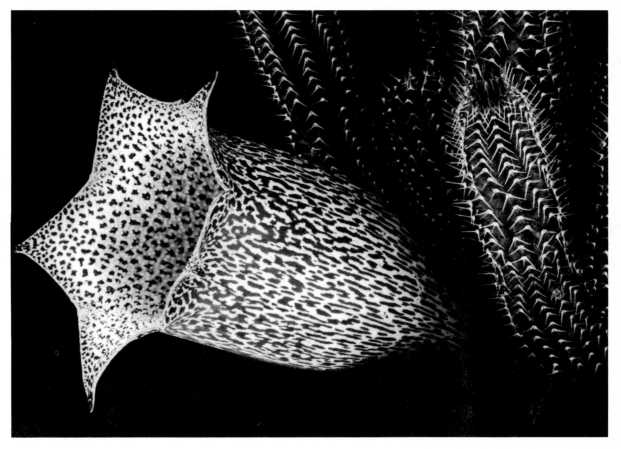

60 *Costus speciosus* with an upright cone-like inflorescence

61 The showy flowers of *Stapelia flavopurpurea* of South Africa, reaching only 3 cm in diameter, resemble stars.
62 *Passiflora racemosa*
63 *Passiflora quadrangularis* is widely distributed in the tropics.

64 *Brachystelma barberiae* uses a curiously divided inflorescence to decoy pollinators.
65 The tiny flowers of the genus *Dorstenia* are embedded in a chocolate-brown pad.

66 The cut-open inflorescence of cuckoopint *(Arum maculatum)* shows, from top to bottom, bristle-like hairs, the male flowers, a second ring of hair, and the female flowers.

67 *Amorphophallus eichleri,* an arum plant of tropical West Africa

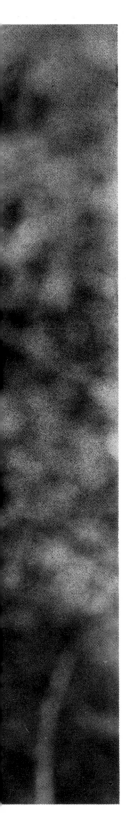

69 Because of its showy flowers the red-brown lady's slipper
(Cypripedium calceolus) has been eradicated in many places.

70 *Aristolochia elegans*
71 *Aristolochia brasiliensis*

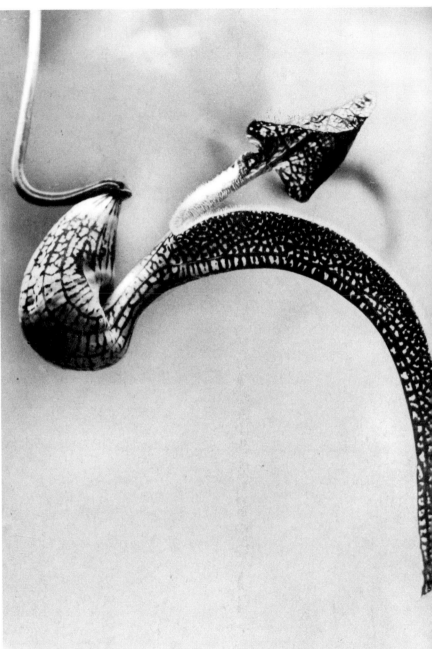

72 By their shapes and colors as well as by a smell of carrion, the flowers of *Ceropegia galeata* decoy pollinators.
73 *Ceropegia distincta* var. *haygarthii*

74 *Angraecum sesquipedale* is the orchid with the longest spur in the plant kingdom.
75 Flowers of the tropical orchid *Stanhopea hernandezii* are spotted and colored like a tiger.

77 Normally flowering plant of the purple orchis *(Orchis purpurea)*
78 Purple orchis in which the flower-stalks are not twisted

79 Flowers of *Epipogium aphyllum* in which the lips are directed upward.

80 In *Cirrhopetalum medusae*, an orchid of the Sunda Islands, the yellow flowers dotted red at their bases are arranged in dense heads and have sepals up to 14 cm long hanging down.

84 The fly orchid *Ophrys insectifera* is widely distributed in
southern and Central Europe: the two leaves resemble a colored fly.
85 The leaves of the spider orchid *(Ophrys sphegodes)* resemble
a spider in shape and texture.

86

86 In its native region, the Mediterranean area, *Ophrys apifera* is pollinated by male longhorn bees.
87 Fruit of the cultivated fig

88 Adam's needle *(Yucca filamentosa)* grows in the deserts of North and Central America.

89 The showy hibiscus flower of *Hibiscus regius* is the state flower of Hawaii.
90 The South American coral tree *(Erythrina crista-galli)* has striking flowers.

91 *Hibiscus schizopetalus* with its bizarre flowers, grows in tropical Africa.

92 Many varieties of the Chinese rose
(Hibiscus rosa-sinensis) were obtained by breeding;
they do not occur in the wild.

93 The bird-of-paradise flower *(Strelitzia reginae)* is a typical bird flower.
94 *Marcgravia evenia* is pollinated by bats.
95 *Datura suaveolens,* a plant with large funnel-shaped flowers, grows in Brazil.
96 The "queen of the night" opens its flowers only at night.

97 *Phalaenopsis amabilis*, an orchid of the Malay Archipelago, is said to flower for a period of four months.
98 *Odontoglossum rossii* flowers for a period of eighty days.

The struggle for light

Halophytes

In marsh and swamp

Carnivorous plants

Specialists in the plant kingdom

Viviparous plants

Plants without roots

Plants without chlorophyll

When speaking of plants in a general way, we think of trees or herbs growing in fields or meadows, in forests, on the wayside or in gardens, or also in flowerpots or window boxes. We consider them as consisting of a well-developed root, a woody or herb-like stalk, leaves, and flowers. For living, the plants require the energy of sun rays, water, and the mineral substances of the soil, as well as the carbon dioxide of the air.

Depending on environmental conditions, a large number of specialized plants have developed which, in their way of life and frequently in their outer shape, differ more or less from the picture we usually form of a plant. The specialists in the plant kingdom have developed this adaptation to the environment in order to occupy habitats where other plants could not exist. In this chapter, we will deal with a few characteristic and interesting representatives of this plant group.

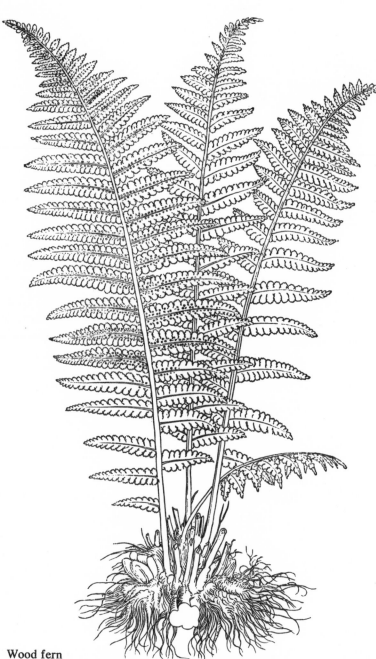

Wood fern

The energy of the sun, sunlight, is one of the most important conditions for the life of green plants, because this energy serves for producing endogenous organic substances from the carbon dioxide of the air and the water of the soil. So it is small wonder that the majority of plants grow toward the sunlight. In some plant communities, especially in dense forests, this is difficult for some plants, so that they have to find certain ways out. In many forests of the temperate zone, for example, the flora on the ground develops in early spring, before the foliage of the deciduous trees closes (Ill. 99). A few ground plants of these forests, similar to the ivy *(Hedera helix)*, are adapted to a lower light intensity.

Different conditions are found in the tropical rain forests where the foliage forms a closed roof throughout the year since there is no pronounced change of the seasons of the year. Forms of adaptation to this environment are, above all, lianas and epiphytes.

In the tropical rain forests lianas are widespread. These plants germinate and root in the soil and frequently form shoots several meters in length. They use these shoots for clinging to other plants, mostly trees, and in this way get to the upper regions of the forest where there is more light. For clinging to the host plants, lianas have modified various organs—leaves, stalks, or roots—so that a distinction can be made between different types of lianas.

In the creepers, leaves, stipules, petioles, stalks or roots have become gripping organs, the tendrils or runners. With them, lianas climb on other plants. Among this group are the passion flowers *(Passiflora)* and the rapidly growing climbing bush *Cissus discolor* (Ill. 100).

The so-called root climbers develop thin roots with which they adhere to cracks in the bark or wind around the stem. Vanilla *(Vanilla planifolia)*, which belongs to the orchids (by the way one of the few orchids not used economically as an ornamental plant), represents this type of liana (Ill. 101). The well-known indoor plants *Monstera* and *Philodendron* grow wild in this way, too.

The spreading climbers avail themselves of spreading branches which grow into the branches of other plants using thorns or prickers as support. The best-known example of this group is the rotan palm (genus *Calamus*) from which the cane so unpopular among former generations of pupils was obtained (Ill. 102). Although this plant belongs to the palms, it is a genuine liana and, having shoots of up to 240 m in length, is one of the longest plants. Spreading climbers also include hop *(Humulus lupulus)*, bittersweet *(Solanum dulcamara)*, and many ramblers, to mention but a few examples.

The convolvulaceous plants form the fourth group of lianas, those that very quickly reach the top of their hosts by their rapidly growing winding shoots. Among them there are, for example, species of beans (genus *Phaseolus*).

A peculiarity of most lianas is the intense water conduction in the shoots. This is necessary to supply a sufficient amount of water to the leaves growing high above the ground in spite of a relatively small stalk diameter. When one cuts a thick liana in the tropical rain forest, water will flow out of the conductive vessels almost as it will from a water pipe; this water can be used as sterile drinking water.

Whereas lianas permanently remain in connection with the forest soil, other herbaceous plants are detached from the ground, taking root in the branches or in branch forks of trees. Such plants are called epiphytes. By the way, many of them are popular indoor plants. Germinating and growing high up in the branches permits these plants to enjoy more favorable light conditions. On the other hand, water supply is difficult. That is why epiphytes depend almost exclusively on rainwater. Their roots serve mainly for clinging to the support, and only in a lesser degree for taking up minerals and water. The plants must be capable of withstanding longer spells of dry weather. Therefore, their modes of adaptation to the adverse water supply conditions are extremely multifarious. The plants have developed quite different devices for economical water consumption and water storage.

An interesting feature of the bromeliads is the cistern-like arrangement of the leathery leaves which collect the rainwater (Ill. 103–105). On the leaves there are many small suction scales through which the water is taken up by the plants. Most of the bromeliads are typical epiphytes. They inhabit branches and grow not only in the tropical rain forest but also in other green forests, savannas, and in desert areas.

Some orchids possess tuber-like leaves as water stores. The fleshy aerial roots of some orchids are capable of sucking up rainwater by means of a tomentose cover, thus placing water at the disposal of the plant (Ill. 106). Another strange feature of the epiphytic habit is that their roots are exposed to light, so that frequently chlorophyll is formed by the root system. Thus, in some species the development of leaves is not necessary and the flowers are put forth directly by the roots.

The epiphytic habitats usually are not particularly rich in nutrients. Due to the poor accumulation of humus in the branch forks, the plants frequently receive only part of the necessary mineral nutrients. For this reason, some epiphytes are in a position to produce humus, such as a few tropical epiphytic ferns which form brown and decaying bay-leaves. Such leaves are developed, for example, by staghorn ferns *(Platycerium)* (Ill. 107). The fertile fronds are the green leaves proper. They are of a bizarre shape, fan-like, forked, and reminiscent of a staghorn. While the young leaves grow upward, the older leaves usually droop. They frequently reach a length of several meters. Bay-leaves are of a roundish shape and are arranged one on top of another like roof tiles. They soon become dry and brown (Ill. 108), and they are decomposed to humus. At the same time, however, they serve as water store and protect the roots from drying up.

In the humus accumulation, and even in certain tuber-like organs of the epiphytes, ants frequently build their nests, as is the case in the so-called ant plants (*Myrmecodia* species) (Ill. 109, 110). Whether this is a true symbiosis has not yet been established.

Usually, epiphytes are distributed by dust-like seeds such as found in orchids, by small spores in ferns, or by birds that feed on berries.

A certain intermediate position between lianas and epiphytes is occupied by the so-called hemi-epiphytes. These plants germinate in virgin forest soils and grow in an upward direction as lianas. When, in the course of time, the lower part of the stem dies, they continue to live as epiphytes. Others germinate as epiphytes in branch forks but later form long roots which grow down to the ground. In the soil, the roots become stronger and the plant displays profuse growth.

A few forms apply their roots to the stems of the host trees; as the roots grow stronger, they slowly strangulate the host tree (Ill. 111). Some fig trees—i.e. certain *Ficus* species—constantly develop new aerial roots at the branches which grow into strong supporting columns (Ill. 112). Such trees frequently reach huge dimensions, having a top circumference of several hundred meters. The largest known specimen of such a banyan tree covers an area of two hectares on an island in the Narbada River in India; the circumference of the crown is 530 m; the diameter, 170 m.

Although the strangulating figs cause the host tree to die slowly, we cannot speak of parasitism in its proper sense, which also applies to lianas and epiphytes. None of these plants extracts any nutrients from its host. We may, however, consider lianas and epiphytes as space parasites, which contend with their host plants for the light.

As we have already discussed, lianas and epiphytes get the required water and light by developing unique forms. Even if there is an abundant water supply, queer habits are developed by plants of which a few will be considered below.

The swamps and bogs, or fens, that have developed in the forest belt of the northern hemisphere, are, probably, the best-known but also most endangered landscape with an abundance of water. This is due mainly to the relatively heavy rainfalls and the relatively low rate of evaporation in these areas. Vast areas of swamps are found in northern Europe, western Siberia, Alaska and Labrador (Ill. 113). These swamps are landscapes of a quite particular charm which normally also excel in producing peculiar plants. Especially the fens have again and again attracted the attention of numerous scientists and many nature lovers. A common feature of all of these regions is a high groundwater level and the occurrence of atypical mosses (*Sphagnum* species), which also play an important role in the formation of fens (Ill. 114).

Atypical mosses possess stalks composed of thick-walled cells which branch out at the top and carry clusters of leaved twigs. Besides narrow green cells for assimilation, they also have colorless cells that serve to store water. Their structure permits them to suck up water several times their own weight like a sponge. That is why swamps have become water reservoirs in nature. One square meter of a cover of atypical mosses stores 6 to 7 liters of water for a period of 10 to 14 days. The lower part of atypical mosses dies away in the course of time, while the upper end continues to grow upward; the parts that die off turn into peat. In this process, the cushions become larger and larger and fuse into one another, thus forming the typical watch-glass shaped surface of the fens. The vertical growth of *Sphagnum* species is anything between 3 and 10 cm a year and all other plants growing in fens must adjust themselves to this growth rate to prevent being overgrown by moss. The annual rate of growth is clearly discernible by the individual annual sprouts.

Fens are developed by alluvial deposits in lakes or by depressions that become boggy because they dam up water or have practically no outlet; examples are the fens situated on

Inula helenium

the crests of the European highlands (Ore Mountains, Thuringian Forest, Harz Mountains). In the Alps fens are still to be found at an altitude of 2,000 m. They are typical in the entire boreal zone, especially in the northwestern regions of the German Federal Republic, in Scandinavia, northeastern Europe, and western Siberia.

The high moisture content and the acid soil poor in nutrients have contributed to the fact that, besides the atypical mosses, a very interesting flora has developed in swamps and bogs. Among other plants, characteristic plants are marsh tea *(Ledum palustre)* (Ill. 115), a plant of the Ericaceae family, the sundew species (genus *Drosera*) which is discussed elsewhere in this book, and the sheathy cotton grass *(Eriophorum vaginatum)* of the sedge family (Ill. 116).

A striking feature of fens is the relatively frequent occurrence of dwarf shrubs of the heather family (Ericaceae) and of the crowberry family (Empetraceae). There are, for example, heather *(Calluna vulgaris)*, the bog bilberrry *(Vaccinium uliginosum)*, the cranberry or marshwort *(Vaccinium oxycoccos)*, and the crowberry *(Empetrum nigrum)*.

Unlike the dwarf shrubs, trees are rarely found in fens and in most cases show a stunted growth. Birches *(Betula pubescens)* and very low-growing pines *(Pinus sylvestris)*, and, in rare cases, mountain pines *(Pinus mugo)* occur in some places.

Swamps and bogs not only are botanical gems and charming landscapes, they are also of economic importance. Formerly, they were dreaded and considered sinister. Today, peat is won from many swamps. Although this raw material is less and less important as a fuel, it still has other uses—e.g. horticulture— and has not been replaced by any other material so far.

But not only the peat-soils are typical vegetation areas with a characteristic flora adapted to an ample supply of water. Bald cypresses *(Taxodium distichum)* (Ill. 117) grow on peat-free muddy soils of low river land and marshy land in southern North America which are flooded by water throughout the year or only during certain periods. An interesting feature of these characteristic plants of the bog forests, which are green in summer, is the fact that the seeds of the trees cannot germinate in water, although the plants require large amounts of water for thriving. These forests are found only where the soil surface dries up regularly every ten to twenty years, and the seedlings reach a height of 30 cm on the unflooded but well-moistened soil. After that, they are capable of standing in mud.

Characteristic of bald cypresses are the pneumatophores which project from the water. They impart a queer aspect to this area. The horizontal roots produce projections, the so-called root knees, at certain intervals, and these are seen above the water surface. These projections are rounded at their tops and serve for oxygen supply and for anchoring the roots to the soil of their boggy habitat. In this way, they also protect the huge trees from windfall. The 40-to-50-m high trees, which are also cultivated in other regions of the earth as park trees, can reach a considerable age. Specimens have been found that were 1,000 to 2,500 years old.

The most curious mode of adaptation to a boggy, frequently water-flooded habitat has been developed by the so-called mangroves.

When coming from the sea and approaching a tropical land with rain forest climate, one will find a so-called tidal forest on flat and muddy coasts and on the estuaries of larger rivers. During high tide only the crowns of the trees project from the water, whereas at low tide one can admire the peculiar aerial and prop roots. Mangroves, as plants growing on the sea shore, tolerate high concentrations of salt because they store a certain amount of common salt in their cells. Close by the trees many pneumatophores grow vertically from the soil or the water. They take up oxygen from the air and conduct it to the root parts embedded in the muddy water. The prop roots of a few species also take up oxygen through pores in the bark. These prop roots frequently develop very queer shapes. Some mangrove trees also develop knee-shaped roots such as are found in bald cypresses. The root tips grow out of the soil, bend downward, grow again into the mud and, after some time, appear again over the surface. This process is repeated several times. The renowned naturalist and advocate of Darwin's theory of evolution, Ernst Haeckel (1834–1919), in his *Indische Reisebriefe*, gives a vivid description of the mangrove vegetation. He writes:

The shore of the island [in the estuary] as well as the banks bordering gardens adjacent to Whist-Bungalow are, like the banks of the river mouth itself, densely grown with the most remarkable mangrove trees, and at my first visit to the nearest neighborhood I had the pleasure of seeing this characteristic and important vegetation form of the tropics in its remarkable land-forming activities.

The trees which are subsumed under the names of mangroves or mangle trees belong to quite different genera and families. An essential common feature of all of them, however, is the peculiar form of growth and the typical physiognomy conditioned by it. The dense, bushy, usually roundish crown of foliage rests on a thick trunk, while this rests on a reversed crown of a bare much-branched root system which emerges directly from the water surfaces to a height of several feet. Between this dense, domeshaped root crown, mud and sand is collected which is deposited by the river along its banks and especially its mouth, and thus the mangrove forest can considerably favor the growth of the land.

In his description, Haeckel points out that mangrove trees in their form and way of life resemble one another closely, but that they are classified in different groups of relations. Although most of them belong to the Myrtales, the genera *Rhizophora, Bruguiera,* and *Veriops* belong to the Rhizophoraceae, the genus *Sonneratia* to the Sonneratiaceae, and the genus *Languncularia* to the Combretaceae, and there are other forms, such as the genus *Xylocarpus* of the order of Rutales and the genus *Avicennia* of the order of Lamiales, which are neither related to each other nor more closely related to the Myrtales. This brief survey also covers the majority of mangrove species, because there are only a few of them.

The mangrove is a typical vegetational form in the tidal zones of the tropics. It thrives, especially in protected bays, lagoons, and river mouths (Ill. 118). The subsoil consists of clayey and sandy mud, which is dry during low tide and flooded at high tide. Another striking feature of the mangrove vegetation is the fact that there is no undergrowth nor are there any lianas such as are found in other humid tropical vegetational forms. Furthermore, epiphytes rarely grow on the trees.

Another peculiarity of mangrove plants, vivipary, is discussed elsewhere in this book.

Conicum maculatum

Halophytes

As we have seen, mangroves are plants that not only are adapted to the life in the tidal zone, but also can tolerate a high salt content in water. For thriving, all plants require a certain amount of salts, which they take up together with water from the soil. An excessive supply, however, will have a lethal effect on the plants. On this basis, a method has been developed for eradicating plant growth from roads, squares, or railway installations, and the like.

In coastal regions and in salty inland places, there are, however, quite a number of plants besides mangrove trees that are not affected by higher salt concentrations, plants which are even dependent on a habitat rich in salts.

The glasswort (Salicornia europaea) of the goosefoot family (Chenopodiaceae) (Ill. 119) is a true pioneering plant that thrives on seashores which are daily flooded with mud-loaded water. The shape of the plant, up to 30 cm in height, is quite remarkable. Thick, fleshy sprouts branch into the air, giving the plant the character of a candelabrum. When the stalks are stepped on, they break, producing a sound reminiscent of breaking glass. The originally dark green shoots, glassy and fleshy, change to a purple-red color in autumn. They seem to be leafless because the leaves are reduced to scales. Thus, the plant is a true trunk succulent which is capable of storing salts in its cellular sap, its mode of adaptation to the environment. As one of the most typical halophytes, it requires a salt content of at least 2.5 to 3 per cent and can even tolerate concentrations of 8 to 12 per cent in water.

The glasswort, flowering from August to November, is the only plant that grows in places with an extremely high salt content where no other plant can thrive; it acts as a natural mud retainer, thus promoting mud deposition. The plant is sensitive to a long-lasting influence of water so that it will die when standing always in water. It thrives best in places where the ground is flooded at certain intervals.

Glasswort is found almost everywhere on the globe. With the exception of Australia, it occurs in all coastal regions of the earth and also in salty inland places. Large areas are still covered with this plant. For example, in the German Federal Republic, it is abundant in Lower Saxony near Münzenberg and in Hesse, and near Mesekenhagen in the vicinity of Greifswald as well as near the saltwater wells of Osterweddingen and Sülldorf near Magdeburg in the German Democratic Republic.

Two other important halophytes, also belonging to the goosefoot family, are the sea blite and the Russian thistle. The sea blite (Suaeda maritima) grows to a height of 10 to 35 cm and has a succulent, branching stalk that bears small oblong leaves. In the axils of the leaves, inconspicuous flowers grow which open from July to September. The whole plant is bald, blue-green in color, and frequently exhibits a reddsih shade. Sea blite is distributed over the same areas as glasswort and occurs in Australia as well. In inland regions, too, sea blite is found in some places.

The Russian thistle (Salsola kali) has a grayish-green color and sometimes shows a reddish shade. The 60-cm-high branching stalk bears small, linear, awl-shaped leaves from whose axils the hermaphrodite flowers are put forth. The Russian thistle, which inhabits seashores, dunes, salty coastal meadows and salty inland places, occurs in South and Central Europe, in North Africa, in Asia Minor and Central Asia, in the Caucasus region, and in Siberia. It was brought by man to North America and New Zealand, where it soon became a bothersome weed.

Artemisia maritima, a plant of the composite family (Asteraceae), is a forb, growing up to 60 cm, which produces dense, tomentose, white sprouts and flowers from August to October. Its main area of distibution is not the coastal regions but inland areas ranging from southern Siberia to the Black Sea. It also occurs on the shores of the Mediterranean Sea, the Atlantic Ocean from South Portugal to Scotland, the North Sea and the western Baltic up to Rügen Island. It is also found in the Central European salty inland regions.

An important halophyte of the leadwort family (Plumbaginaceae) is the sea pink (Limonium vulgare). The up to 15-cm-long, obovoid, somewhat fleshy leaves form a basilar rosette from which the up to 50-cm-high stalk rises, bearing blue-violet flowers. The flowering time extends from July to September. The sea pink inhabits salt meadows and muddy areas that are more or less flooded during high tides on flat seashores; it is a gregarious plant. Its area of distribution includes the Atlantic coasts of North America and Europe, and the coasts of the North Sea, the western Baltic, and the Mediterranean. Related species are found in the salt steppes of southeastern Europe, the southern Soviet Union, Asia Minor, and on the South American shores. The sea pink does not store salt in its cellular sap, but in the leaves and the stalks are specialized cells which excrete excessive amounts of salts.

Tamarisk

The study of the natural circulation of the substances has made evident the fact that numerous animals feed on plants. Furthermore, as plants are the only organisms that build up organic substances from inorganic matter, in the final analysis all living organisms are dependent upon plants as nourishment, whether directly or indirectly. Therefore, it is small wonder that the reverse process—i.e. plants feeding on animals—attracts the special attention of scientists and laymen ever again.

Apart from carbon taken up by plants through their leaves from the carbon dioxide of the air, they take up water through their roots, and with it all other nutrients in the form of dissolved salts from the soil. These salts contain inorganic substances such as nitrogen, sulphur, phosphorus, potassium, calcium, magnesium, and many others that are required for building up the organism and for maintaining the vital functions. An exception are carnivorous or insectivorous plants, also called insectivores. These are plants which possess a device for catching small living animals, usually insects, are capable of decomposing proteins by the secretion of enzymes or with the help of bacteria, and of ingesting the decomposition products of animal protein through special digestive glands. In this process, only the empty solid chitins of insects are left behind, lying on the catching organs or in the tubes or cans of the plants.

The biological significance of this way of nourishing mainly is in the acquisition of nitrogen, which normally is taken by plants from the soil. Most of the insectivorous plants, however, grow in habitats very poor in nitrogen—e.g. in fens—and therefore have to rely on an additional source of nitrogen. This source of nitrogen is the protein of the caught animals.

About 450 species of insectivorous plants occur on the globe. They are divided into two large groups, irrespective of the particular forms of their catching mechanisms. The closely related orders of Saxifragales and Sarraceniales include the genera *Nepenthes, Dionaea, Sarracenia* and *Drosera,* while the genera *Utricularia* and *Pinguicula* belong to the Lentibulariaceae family of the order of Scrophulariales. The species of these genera are discussed later in this chapter.

The most important peculiarity of all insectivorous plants is the various catching devices which have developed through

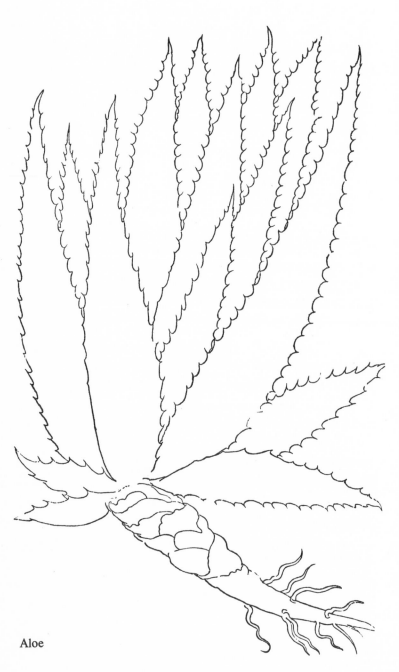

Aloe

modification of individual organs. They include simple sticking traps, slide traps that operate on the pitfall principle, and very complicated flap traps that respond with rapid motions to certain stimuli.

The oldest known description of an insectivorous plant, probably the first one, was given by the Englishman John Ellis in 1769, who wrote a letter including a description of a Venus's-flytrap *(Dionaea muscipula)* to the Swedish botanist Carl von Linné, the founder of the first plant system and of the binary nomenclature in biology. Few people, however, believed in the information given by Ellis, because the existence of carnivorous plants was thought to be improbable. Yet soon, other botanists began to take an interest in these curious plants. For five years, Darwin conducted experiments with sundew *(Drosera)* and butterwort *(Pinguicula)*. In 1875, he published his famous book *Insectivorous Plants*, which is still considered a standard work in this field.

The simplest trapping device among insectivores is the so-called sticking trap found in the sundew species (genus *Drosera)* (Ill. 120). About ninety sundew species are known to be widely distributed over the earth. The main distribution areas are South America, especially Brazil; South Africa; Australia; and New Zealand. Three species occur in Central Europe, and all are rosette plants, growing mostly in bogs. These are *Drosera rotundifolia, Drosera anglica,* and *Drosera intermedia.* Although they reach a maximum height of only 20 cm and therefore are easily overlooked, all species are protected by nature preservation regulations in most countries, because particularly keen "nature lovers" like to gather these interesting plants, thus decimating the population.

The leaves of sundew have been modified into catching organs. Their blades are covered by stalked glandular hairs, the tentacles, which serve for catching, holding, and digesting the insects. These tentacles are distributed over the whole blade, while the length of the stalks decreases toward the center. At the head-shaped tips of the tentacles, the plant secretes a clear, bright, slimy drop that smells like honey and attracts insects, usually small flies. The victim sticks to these droplets and, stimulated by the contact, all of the plant's tentacles bend over it, so that it is completely covered with slime and finally suffocated. The conduction of stimulation in the leaves proceeds at a speed of about 8 mm per second. After the suffocation of the creature, the plant secretes protein-decomposing enzymes that disintegrate the animal protein so that it can be ingested by the plant. Shortly afterward, the tentacles are erected again, ready for catching new victims; only the empty chitin of the insect is left on the leaf surface, and that is soon blown away by the wind.

Although the two genera are not closely related, butterwort (genus *Pinguicula)* also develops sticking traps. There are about 40 species of butterwort, which mainly occur in North and Central America but also in Europe and northern Asia. Steepwort *(Pinguicula vulgaris)* (Ill. 121) which has violet flowers, and *Pinguicula alpina,* with white flowers, grow in Central Europe. Botanically, it is of particular interest that butterwort is included in the dicotyledonous plants, though only one cotyledon is formed in seed germination.

Butterwort also catches insects by means of leaves arranged like a rosette on which there are two kinds of glands—petiolate, sixteen-cell glands and nonpetiolate, eight-cell glands. About 10,000 of such glands were counted on a leaf surface of one square centimeter. When an insect alights on the leaf, it gets stuck to the slime of the petiolate glands, which immediately secrete more slime, enveloping the animal. In addition, the edge of the leaf is rolled up so that the insect finds itself in a slime-filled tube. Then the nonpetiolate glands secrete a protein-decomposing enzyme so that the insect is digested. After one to three days, the leaf unrolls again. This reaction can be carried out by a leaf only two or three times; then it dies.

Whereas sundew and butterwort possess sticking traps, the plants described below have slide or pot traps formed of modified leaves or parts of leaves.

In damp and boggy areas of eastern North America, the sidesaddle flower *(Sarracenia purpurea)* (Ill. 124) grows. The leaves of this plant reach a length of up to 75 cm and are modified into pitfalls that sometimes show a bizarre trumpet-like or pitcher-like form. The top of the tubular leaf takes the form of an umbrella or is provided with a cover. Insects are attracted by the remarkable size and the gay purple color of the upper leaf section and by nectar glands and thus are caused to slip down into the interior of the tubular leaves. These tubes can be divided into different zones. The upper end is covered with honey glands and long rigid hairs directed downward. Below this is a sliding zone that leads into a zone provided with hairs arranged in the form of combs directed downward. Finally, there is the zone containing the digestive glands. At the bottom of the tube, water, dissolved enzymes, and acids are collected which drown the caught insects and decompose them. Besides these species, eight other species occur in North America which mainly differ in size and shape of the tubular leaves. As several of them have the same area of distribution, a large number of hybrids have developed.

Pitcher plants (genus *Nepenthes)* seem to be perfect in specialization; of them, about seventy species occur in the tropical rain forests of Asia, Madagascar, Indonesia, and Australia (Ill. 122). They grow mainly as epiphytes on trees or among decaying leaves on the ground. A creeping rhizome produces a climbing stem that bears large leaves and runners. A pitcher provided with an open immovable cover is suspended on many leaves. The pitchers are modified leaf blades, and the leaf forms the bottom of the pitcher. In between, there is the petiole, which has changed into a runner.

Usually, half the pitcher is filled with water which, in spite of the dead insects lying at the bottom, is normally clearer and more suitable for drinking than the water in the surrounding bog. People have drunk the water of these pitcher plants and did not come to harm; this is why such plants are also known as "hunter's cup" in some regions.

Attracted by honey, insects creep into the interior of the pitcher which is divided into several zones, as is the case in tubular plants. The smooth inner walls of the pitchers offer no hold to the insects, so unavoidably they fall into the water and drown. An insect is digested within five to eight hours, only its chitinous integument being left. Up to 6,000 digestive glands have been counted on one square centimeter of the interior area of such a pitcher.

The pitchers have quite different shapes, and their colors vary between green and a deep brown-red. Their size is 3 to 50 cm. The display houses of botanical gardens primarily show horticultural breeds such as *Nepenthes x mixta*, which develops beautifully patterned pitchers growing up to 30 cm in length (Ill. 123). Cultivation and care of this plant are easy.

More advanced methods of trapping are used by the bladderwort and Venus's-flytrap, which have what are known as "flap traps."

There are 275 species of bladderwort (genus *Utricularia*) distributed all over the globe but found primarily in the tropics. Six species grow in Central Europe. The best-known species is the common bladderwort *(Utricularia vulgaris)*. Most of the species are submerged aquatic plants; a few representatives also grow in bogs or in virgin forests as epiphytes.

The blister-like catching devices, being provided with feeler bristles and closed by a cover, are located in the axils of the finely divided leaves. The blisters contain air. When a small aquatic animal (crab, rotifer, etc.) touches the feeler bristles, the cover opens inward and with the inflowing water the animal is swept into the blister. Inside the blister, the animal is disintegrated by enzymes secreted by the plant and probably also by bacteria contained in the blister. The particles of animal protein are thus digested by specialized cells.

Darwin called the Venus's-flytrap *(Dionaea muscipula)* one of the most wonderful plants of the world (Ill. 125) because of the remarkable rapidity of its motions. The plant, the only species of the genus, which grows in bogs in the Carolinas of North America, has thick basilar leaves resembling an open shell and being borne on a petiole broadened in the form of a leaf. The leaf surfaces are provided with rigid bristles on their edges, each leaf having three feeler hairs on the inner side. The two halves of the leaf are connected by a joint along the central axis.

An insect touching the feeler hairs causes the halves of the leaf to come together immediately. This movement is initiated by pressure variations in the cells. During the closing motion, the tooth-like elements of the two leaf halves get hooked up so that even relatively large insects cannot escape. Digestion

is effected by digestive glands on the inner surface of the leaf. The digestion process takes one to two weeks in the Venus's-flytrap. Then the leaf opens and is again ready for catching. Usually, a leaf can carry out only two catching processes, then it soon dies, as is the case with butterwort.

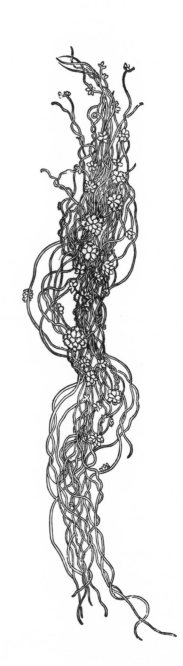

Flax

Higher plants multiply by seeds produced as the result of fertilization. These seeds get detached from the parent plant and, after a fairly long period of dormancy, they germinate in the soil. That is how new plants grow. Exceptions are multiplication by offshoots, by division, or other processes.

In contrast to plants, many animals bring forth living young, as is the case with almost all mammals. This is called viviparity. Viviparity, however, is also found in plants, but in rare cases only. Many mangrove trees do not scatter seeds; the seeds germinate on the parent plants and develop into relatively large sprouts which swing to and fro suspended on the branches like inverted candles (Ill. 126). When they have reached a length of 60 to 80 cm, they are fairly heavy and fall from the parent plant and penetrate into the mud. After a short time, they form long roots. This viviparity is a true adaptation to the living conditions of the mangroves. In soil exposed to tidal flooding, small seeds could not survive; they would lack oxygen and would be washed away by the motion of the water.

While mangroves are true viviparous trees, a false viviparity in several plants exists too. Instead of flowers, these plants develop leaf sprouts, or breeding organs are formed on other parts of the plant which fall off and become new plants. One of the best-known examples is the *Bryophyllum* (genus *Kalanchoe*) (Ill. 127), which is widely kept as an indoor plant. On the surface or edges of the leaves, small newly bred plants develop. These easily break off or fall off, and most have small roots, so that when they are in contact with the soil, they soon take root, forming independent plants. The bryophyllum also puts forth showy terminal inflorescences that produce large amounts of seeds, thus contributing to the preservation and wide distribution of the species.

Johann Wolfgang von Goethe (1749–1832) also took a lively interest in the bryophyllum. For many years he observed this interesting plant not only in the plant houses of Belvedere near Weimar but also kept a few specimens in his house. Several of his notes on the bryophyllum have become known. In 1830, he sent a seedling of the plant to Marianne von Willemer in Frankfort on the Main with a letter in which he wrote:

As from a leaf
countless other branches weave,
May one love secure
for you a thousand pleasures more.

Kalanchoe tubiflora, depicted in this book, is native to southern Madagascar and can be easily kept in the open during summer and in a cool, well-lighted room during winter. Numerous flowers develop in January and February.

False viviparity is also found in other plants native to the temperate zones where it frequently occurs as a substitute for the nonoccurrence of fruit and seed formation. For example, in orchard grass *(Dactylis glomerata)* and in a meadow grass *(Poa alpina)* young sprouts are formed in the place of flowers. In knotweed *(Polygonum viviparum)* small knots are put forth in the inflorescence instead of flowers, and these produce leaves, which, when they fall off the parent plant, immediately take root. Various species of garlic, leek, and the like, including allium *(Allium paradoxum)* and sandleek *(Allium scodoprasum),* have bulbils as well as genuine flowers in their inflorescences. Such bulbils are also found in the inflorescences of rockfoil *(Saxifraga nivalis).* But these plants put forth small daughter plants not only in the region of the flower and the inflorescence. In toothwort *(Dentaria bulbifera)* and in pilewort *(Ranunculus ficaria)* bulbils are formed in the leaf axils and in sea holly *(Eryngium viviparum)* small sprouts are formed in the axils.

Plants without roots

In general, all of the higher plants possess roots which serve for holding the plant in the earth, for supplying it with water containing dissolved mineral substances, and also for storing water and nutrients. Moreover, roots can be modified into pneumatophores, runners, or prop roots, and even into assimilation organs as, for example, in bald cypresses, lianas, and in some orchids.

In adaptation to a certain mode of life, the roots of several plants are completely stunted. This phenomenon is found in many aquatic plants that are not rooted to the bottom and that take up the water nutrients through their whole surface. Such species no longer require roots. One example is the bladderwort, which has already been discussed.

In the bromeliad family, occurring in the American tropics, there are a few plants without roots which, however, do not live in water. In the dry forests of South and Central America, the bromeliad *Tillandsia usneoides*, a rootless plant (Ill. 128, 129), grows. This epiphyte, also known as Louisiana moss or Spanish moss, has the appearance of beard moss. It hangs down from the branches of trees, frequently in such large quantities that it forms a long and dense curtain. Often it completely covers telegraph wires.

The plant takes water through its leaves, sometimes in quantities amounting to six to ten times its dry weight. As a consequence, the epiphytes become very heavy, causing telegraph wires or branches to break. Although *Tillandsia usneoides* is the smallest and most finely structured species of its genus, it is the most dreaded one in populated areas. Birds or the wind distribute this interesting plant by carrying seed or small parts of the sprouts. The wiry and winding sprouts and the fine twisted leaves anchor in the bark of trees. Every detached piece is capable of growing independently. For its further development, it does not require any roots, only a certain degree of humidity. Humidity is vital to the plant, for it takes in water with the help of small suction scales distributed over the whole plant body.

Tillandsia recurvata (Ill. 130) also thrives without roots, though not in the form of such long and voluminous beard tongues. With regard to vitality, however, it is not inferior to *Tillandsia usneoides*, and it can also occupy telegraph wires. The *Tillandsia* species, however, are not the only terrestrial plants without roots. In Europe, too, rootless plants are found.

In winter, when the deciduous trees have shed their leaves in North and Central Europe, one can see the mistletoe *(Viscum album)*, a parasite on various trees occurring in the whole deciduous forest region of Europe and North Asia (Ill. 132). Poplars, acorns, limes, false acacias, and apple trees are commonly infested by mistletoes. In addition, however, there is the mistletoe *Viscum laxum*, which grows on pines and firs, and the mistletoe *Loranthus europaeus* which lives only on oaks. As the mistletoe has green leaves, hence can take carbon from the air, it is only a hemiparasite. The most remarkable feature of this plant is its ability to live as a parasite on plants of its own kind. So we can sometimes see small "super"-parasites on larger clusters of mistletoe. The sticky seeds get on trees through birds. In these trees, the seeds germinate, though genuine roots are not formed. In place of roots the plants develop so-called bark suckers, which penetrate the bark of the host plant and drive tapering shoots into the woody part of the host from which they take water containing dissolved substances. The articulate stalks are brittle and branched. At their ends, they bear two leathery leaves growing opposite to each other, having a pale green or gold-green color.

The mistletoe is a dioecious plant that has either only male or only female flowers. It is interesting to note that in nature the number of female plants is three times that of male plants. The reason for this phenomenon is not fully understood. For pollination, which is effected by small flies, the small number of male plants seems to suffice, however.

The small female flowers of yellow color appearing in spring develop, in the course of the year, into white transparent berries the size of a pea, which are filled with a sticky sap (Ill. 131). They are distributed by certain birds, above all the mistle thrush and the waxwing.

Both in antiquity and in Germanic mythology, the mistletoe plays an important part and for a very long time, this evergreen plant has had a symbolic significance in England and in parts of France at Christmastime. It is hung up at the ceiling or over the door, and according to an old custom, a couple meeting under the mistletoe may kiss. Formerly, mistletoe preparations were used as a hypotensive agent, a medicine for lowering blood pressure.

It is generally believed that plants have green leaves that serve for building up organic substances from inorganic substances by means of the chlorophyll contained in them. There are, however, plants that contain no chlorophyll. Therefore, they must obtain organic substances in a different manner. Frequently, plants without chlorophyll belong to species of low-growing plants. The best-known representatives are mushrooms. But even among the flowering plants, there are a few species that do not possess any chlorophyll.

In various European forests, a few of these interesting plants grow. They are conspicuous because of their pale color and particular mode of gaining nourishment. Such plants, which live on dead or decaying organic matter, are called saprophytes. They draw nutrients from the substrate on which they grow, usually humus. As they do not require light for photosynthesis, they are frequently found in the darkest places in forests, and they form a striking contrast to the green plants around them.

One of the saprophytes is the bird's-nest orchid *(Neottia nidus-avis)*. It has a 20-to-40-cm-high inflorescence of a leather-yellow to light-blue color (Ill. 135). The relatively large flowers are yellowish and have a faint smell of honey. The bird's-nest orchid, however, does not only live saprophytically but occasionally also as a parasite; that is to say, it draws nutrients from other plants. It always occurs together with a root mushroom in a so-called symbiosis. Both orchid and mushroom are dependent upon this symbiosis. This mushroom-root symbiosis is of importance to the bird's-nest orchid, in two respects, applying to many other orchids, too. First, the mushroom exerts a germinating stimulus on the seed as without this mushroom the seed is not capable of germination. Second, the mushroom feeds the young plant with nutrients necessary for its further development which are taken from the soil by the mushroom.

The coral-root *(Corallorhiza trifida)*, an orchid of the beech and fir forests of Europe, Siberia, and North America, also lives as a saprophyte and parasite (Ill. 134). The stalk of the slender and delicate plant, growing 6 to 20 cm high, is of a yellowish-green color and is capable of assimilating to a certain degree. The small flowers have a white trilobate lip covered with red dots surrounded by yellowish petals. These two orchids

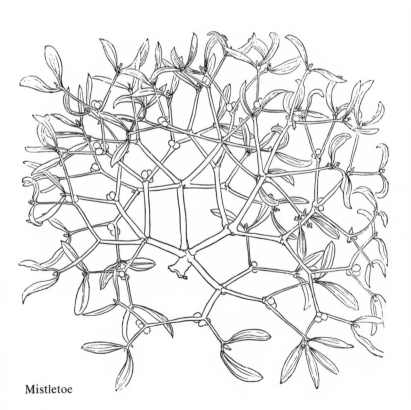

Mistletoe

flower from May to June and are frequently found in communities.

A similar mode of life distinguishes the Indian pipe *(Monotropa hypopitys)*, which occurs in damp deciduous and coniferous forests of the entire northern hemisphere outside the tropics. The pale, brown or wax-yellow plant, rarely showing a pink or purple shade, reaches a height of 10 to 30 cm. Its stalk is covered with scale leaves, densely growing and reaching a length of 1 to 1.5 cm. Its flowers unfold from June to August. In a symbiosis with root mushrooms, it grows on mild humus as both saprophyte and parasite. Among the plants without chlorophyll, there are, however, also representatives that do not draw their nutrients from mushrooms but from higher plants; that is to say, they lead a true parasitic life.

Representatives of the silkweed plants *(Cuscuta)* consist only of winding stalks and flowers; they are distributed all over the world. Their stalks wind around their host plants, from which they take all the nutrients they require, driving special sucking outgrowths, the so-called haustoria, into the hosts (Ill. 133). Silkweed is found on trees and, more frequently, on herbs. As various species such as *Cuscuta trifolii, Cuscuta europaea* and *Cuscuta epilinum* also infest certain cultivated plants—e.g. clover, hop, and flax—they occasionally can cause considerable damage. These plants are also known as boxthorn or matrimony vine.

Plants of the broomrape family (Orobanchaceae), of which the most widely distributed genus is broomrape *(Orobanche)*, are also parasites on higher plants but cause less damage. They occur on a wide range of wild herbs, especially on plants of the composite family, but rarely on cultivated plants. These are chlorophyll-free herbs with upright stalks covered with scale leaves. They reach a height of up to 60 cm, having a yellowish, brown, or violet color. Most species occur in Europe and Asia; only a few are native to North Africa and North America. The broomrape plants take their nutrients from the roots of their hosts into which they drive their haustoria. They themselves do not develop true roots. Close relatives of broomrape plants are plants of the figwort family (Scrophulariaceae), to which belong parasites, especially hemiparasites.

The resistant chlorophyll-free toothwort *(Lathraea squamaria)* is a true parasitic plant reaching a height between 10 and 25 cm and occurring in Europe and in the temperate regions of Asia up to the Himalayas (Ill. 139). This plant has attracted the attention of a large number of botanists because of its odd appearance and peculiar mode of life. Mostly it is a subterranean plant. In spring, however, the racemes which are fecund, rich in sap, and faintly reddish, appear above the ground. A richly branched rhizome, consisting of a main root with a large number of side-roots, grows underground. The side roots envelop the root system of the host plant, drawing water and nutrients from it through their haustoria. Toothwort is a parasite especially of trees and shrubs such as alders, hazels, beeches, hornbeams, oaks, elms, and many park trees.

Besides the true parasitic toothwort, a number of hemiparasites are included in the figwort family; one hemiparasite, the mistletoe, belonging to a quite different group of relatives, has already been discussed. Figwort plants, however, do not live in the crowns of trees but draw their nutrients from the roots of their host plants in a manner similar to the toothwort. Among the large number of genera are the rattle *(Rhinanthus)*, eyebright *(Euphrasia)* (Ill. 138), lousewort *(Pedicularis)* (Ill. 136), and cowweed *(Melampyrum)* (Ill. 137), to mention but a few examples. Although all of them have green leaves, they remain small and do not flower without connection with host plants. Most of them are parasites of grasses, plants of the sedge family and a few other meadow plants. Especially the large rattle *(Rhinanthus serotinus)* may sometimes become a nuisance in cereal crops (Ill. 140). As to the selection of host plants, some cowweed species form an exception, growing as parasites on forest trees and bilberries.

100 *Cissus discolor* is a rapidly growing climber which soon reaches the tops of virgin forest trees.
101 With its thin roots vanilla *(Vanilla planifolia)* clings to the trunks of virgin forest trees.

102 Rotan palms (genus *Calamus*) produce shoots having a length of up to 240 m.
103 *Tillandsia prodijosa* is an epiphytic bromeliad.

104 Rainwater is collected in the cistern-like leaf arrangement of bromeliads.

105 Leaf scales of the bromeliad *Tillandsia hildae* collect water (electron-microscopic photograph).
106 Epiphytic orchid with aerial roots which can take up rainwater.

107 Staghorn ferns *(Platycerium)* on a baobab
108 The lobed fronds of staghorn fern

109 An ant plant *(Myrmecodia echinata)*
110 Cut tuber of an ant plant with cavities in which ants live

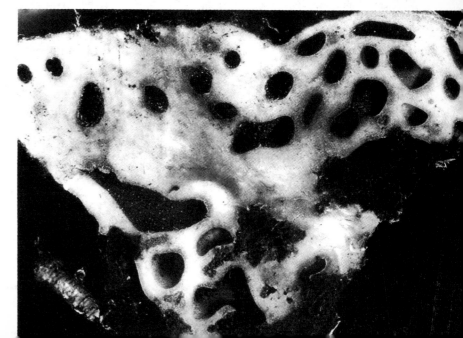

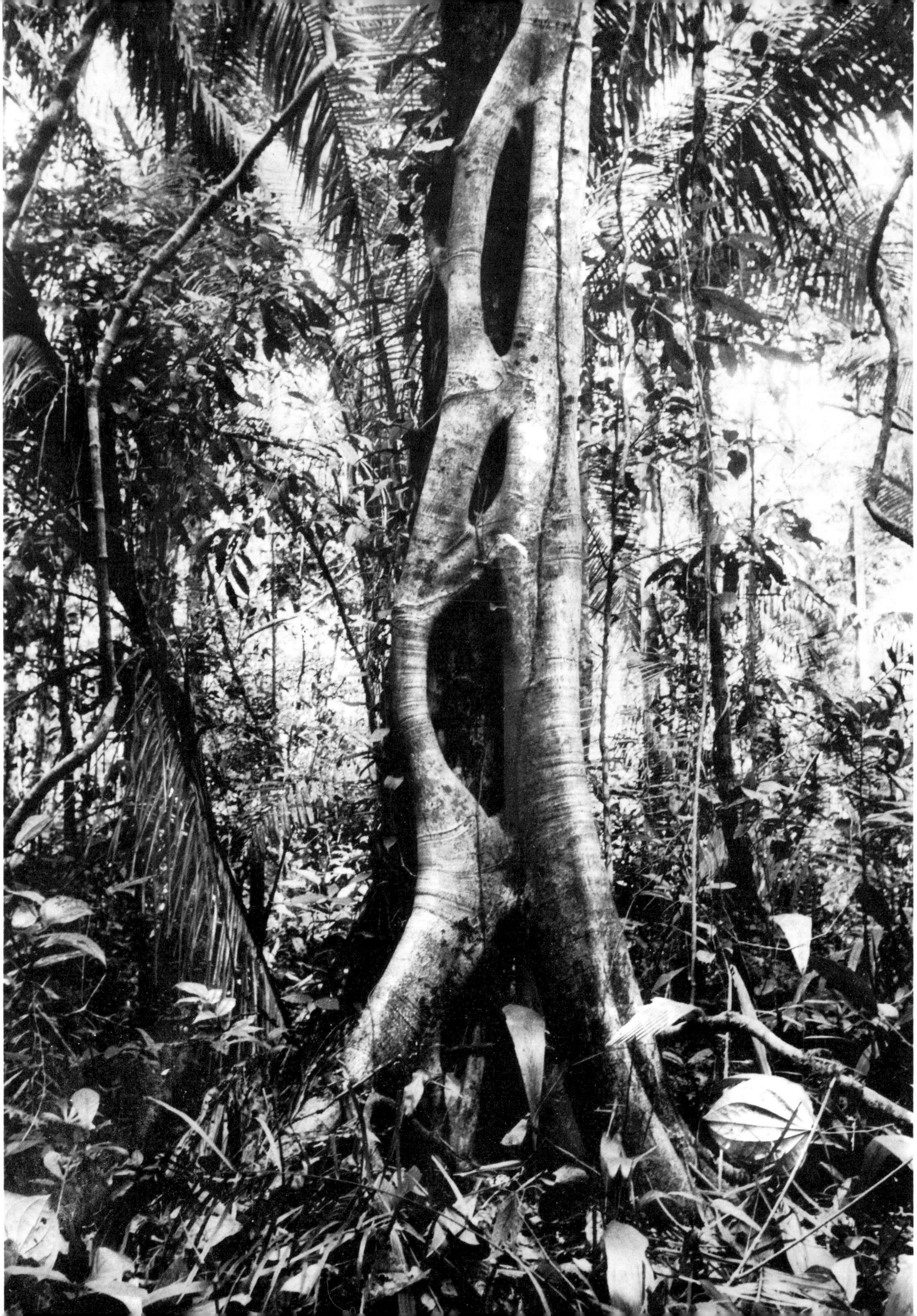

111 *Ficus bengalensis* on a carrying tree in a virgin forest of southern Peru
112 Old, big specimen of *Ficus bengalensis* with many aerial roots growing on Mauritius

113 A typical high moor in northern Europe with a flora comprising many species

114 Most typical of the high moors of northern Europe and Asia
are *Sphagnum* mosses.

119

115 *Ledum palustre,* also known as marsh tea
116 Cotton-grass *(Eriophorum vaginatum)*
117 Bald cypress root knee *(Taxodium distichum)*

118 Various developmental stages of the mangrove plant *Rhizophora mangle* in Cuba. The respiratory roots of *Avicennia nitida* can be seen in the background.

120 Leaf rosette of a sundew plant *(Drosera capensis)* with stalked tentacles

121 Butterwort *(Pinguicula vulgaris)* catches insects with small glandular hairs on the leaves.

122 Bog vegetation, including the pitcher plant
Nepenthes madagascariensis, near Mantanino in Madagascar

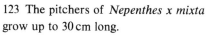

123 The pitchers of *Nepenthes x mixta*
grow up to 30 cm long.
124 Tubular leaves of *Sarracenia purpurea*
125 The leaves of Venus's-flytrap *(Dionaea muscipula)*
come together to trap insects.

126

126 Young germs on the branches of the mangrove plant
Rhizophora conjugata in Madagascar
127 Bryophyllum with germ-buds

128 Individual plants of *Tillandsia usneoides*
129 A tree densely covered with *Tillandsia usneoides*
in the Mexican fog forest

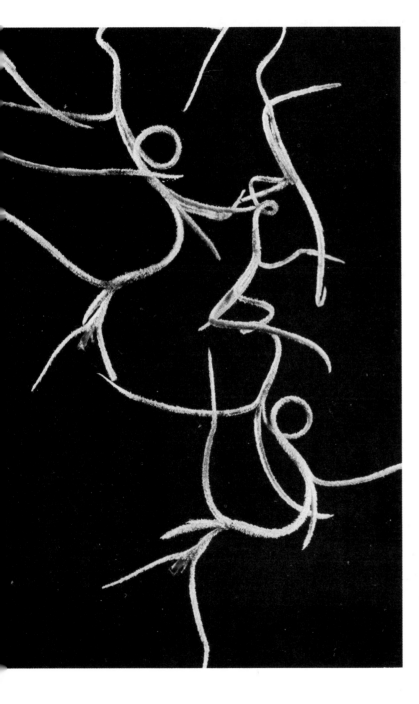

130 *Tillandsia recurvata* growing on an electric wire

131 Fruit of mistletoe *(Viscum album)*

132 The mistletoe *(Viscum album)* is particularly well discernible on the leafless trees in winter.

130 133 Dodders (genus *Cuscuta)* sponge on a wide range of hosts.
134 The coral root *(Corallorhiza trifida)*
135 Bird's-nest orchid *(Neottia nidus-avis)*

136 The lousewort *(Pedicularis palustris)* is a partially parasitic plant which inhabits moist places.
137 The cowweed *(Melampyrum nemorosum)* attracts attention especially because of its violet-yellow inflorescences.

138 The common eyebright *(Euphrasia officinalis)* is a semiparasite.

139 The pink inflorescences of toothwort *(Lathraea squamaria)* can be found in various places in spring.
140 The rattle *(Rhinanthus serotinus)* is a partially parasitic herb.

Plants and superstition

The wonder-working mandrake   The gist of the matter

Magic plants –
witch plants

Drugs and poisons

From arrow poison to drug

Ginseng between superstition and reality

The road to "Paradise"

Vervain

Plants are used by man not only for purely economic purposes, as food or raw materials, they also play an important role in customs. In this connection, we should like to mention that flowers are presented on the occasion of happy and sad events in almost all parts of the world.

Superstitious ideas have been, and still are, closely associated with plants. Formerly they were widespread, but even today there are people who have not yet overcome them completely. For example, many people are pleased to find a four-leaved clover *(Trifolium repens),* considering it a token bringing good luck. There even is an indoor plant that is called four-leaved clover *(Oxalis deppei).* It is, however, not related to the white clover growing in meadows.

These remains of superstition, however, are quite harmless as compared to the ideas in the Middle Ages or in ancient times.

For hundreds of years, people have attributed witchcraft to certain plants. The close association of this belief with plants is reflected in many traditions and legends; proof of this is also given by a number of plant names such as witches'-broom-stick, enchanter's nightshade, witch-meal, witches'-ring, devil's-milk, and others. The origin of this superstition is our ancestors' belief in demons. Now let us have a look at some of these witch plants.

Garlic *(Allium sativum)* was said to protect human beings from witchcraft and sorcery. In some regions it was added to animal foods in order to protect the animals against evil. Sweet woodruff *(Galium odoratum)* also served for warding off evil spirits. Caraway *(Carum carvi)* was likewise cultivated for that purpose. Seeds of peony *(Paeonia officinalis)* were used in particular to protect children from witches and devils (Ill. 147).

It was, however, not only necessary to be protected from witches. One had also to distinguish them, as they did not differ from other people in appearance. This problem was solved by a cross of ground ivy *(Glechoma hederacea),* centaury *(Centaurium minus)* or lovage *(Levisticum officinale).* Mother-of-thyme *(Thymus serpyllum)* was held to bring good luck and was also useful for exorcism or healing complaints caused by

the devil. "Botanists" of former times could not bring themselves to pass by Saint-John's-wort *(Hypericum perforatum),* which secretes a red-colored substance and whose leaves appear to be perforated (Ill. 141). A small bunch of this plant was believed to ward off a lightning stroke.

Because of the peculiar mode of life of mistletoe, which has already been discussed, witchcraft was also ascribed to this plant *(Viscum).* Hung up in the house or the stable, it was said to protect man and animal.

Plants having magic powers were not only used to drive away witches and devils; witches themselves needed them to exercise their craft and do mischievous things. They made an ointment of henbane *(Hyoscyamus niger)* and thorn apple *(Datura stramonium),* deadly nightshade *(Atropa bella-donna),* monkshood *(Aconitum napellus)* (Ill. 142) and vervain *(Verbena officinalis)* which made them invisible and took them with lightning speed to the witches' Sabbath on the Blocksberg. We shall come back again to this witch's ointment.

All plants discussed so far are clearly definable, but this is not the case with some others, although they are no less known.

The poet of classic antiquity, Homer, in his *Odyssey* describes the herb moly. As is well known, the companions of the Greek hero were caught in the spells of Circe, a witch who transformed them into pigs. When Odysseus set out to liberate his friends, he was warned by Hermes, messenger of the gods, who gave him the herb moly which would protect him from the fatal drink. Unfortunately, Homer says only that the plant has a black root and mild-white blossoms. For this reason, it has unfortunately not been possible to identify the herb moly although Theophrastus *(c.* 372–287 B.C.) and Pliny (A.D. 79), Linné, and even botanists of recent times have made great efforts to find it. Probably this plant is purely mythological.

Jump root and magic flower play a role in many fairy tales or legends ascribing to them curious magical powers. A person possessing the magic flower becomes invisible to other people and easily finds access to caves where treasures are hidden. Unfortunately, these fairy tales in most cases do not have happy endings because the finder of the magic flower loses it again.

The magic flower is said to be a blue or white blossom which might resemble a chicory or a lily. Up to now, scientists have not been able to ascertain more details.

Information about the jump root is also incomplete. It likewise was said to render its owner invisible and to be capable of opening closed doors. To get this flower, one had to stop up the nest of a woodpecker or a hoopoe in such a manner that the birds could not get access to their hatches. In this case, the animal brought the desired root, opened the nest, and dropped the root, which thus got into the hands of the happy finder.

Strange powers were also ascribed to fern seeds. They were said to make people invisible, invulnerable and to help them to discover treasures. It was difficult, however, to collect fern seeds because they could be found only on St. John's night (24 June) and, in addition, could only be won with the help of particularly dangerous magical invocations. Today, everybody knows that ferns do not produce seeds. Just this fact may have been conducive to the development of superstition because people used to believe that all plants develop seeds. Why should the large and conspicuous ferns be an exception?

No other plant enjoyed such a high esteem in the superstition of former times in European and Oriental regions as the mandrake *(Mandragora vernalis)* (Ill. 143). Even the ancient Egyptians must have known it, as a pictorial representation on a tomb wall erected in 1350 B.C. shows. Although the Greek botanist Theophrastus reported on the mandrake and criticized the humbug of magic rites during the digging for the root, fairy tales about its witchcraft again and again appeared in the following centuries. The Jewish historian Josephus Flavius (A.D. 37–95) reported on this in his *History of the Jewish War.*

The valley enclosing the town of Machaerus [to the east of the Dead Sea] on its northern side is called Baara and produces a marvelous root of the same name. It is flaming red and emits red rays in the evening; it is very difficult to uproot it because it evades any approaching person and will only be stationary if one pours urine and blood on it. Even then death is certain when touching it unless one carries the entire root away in one's hand. But it can be obtained in

Digging-out of mandrake in a medieval representation

another way. One digs out the earth around it so that only a small remaining part of the root is still hidden. Then one ties a dog to the root and when the dog wants to follow his master, he will tear out the root but will die immediately as a representative victim of the man who wants the plant. Once one has got possession of it, there is no longer any danger. One takes such great pains to get it because of the following properties of this plant: Demons, that is to say, evil spirits of bad people, which get into living people and kill them unless help is immediately available, will be driven away as soon as this plant is brought near a sick person.

Moreover, it is said to be a remedy against epilepsy and eye diseases, effective as an aphrodisiac, and bring hidden treasures to light. Goethe refers to the latter property when he makes Mephistopheles say in the second part of *Faust:*

So they stand around and gape,
Unable, they, their find to trow.
Mandrake! the judgment one doth shape,
The Dark One! breathes the other now.

Now what can really be said about this plant? The mandrake is one of the nightshade plants, which possesses a strikingly branched storage root, which with some imagination may be considered as a human figure. It is distributed from the Mediterranean region to Southwest Asia up to the Himalayas and flowers from the end of February to the beginning of April, exhibiting a large number of inflorescences close above the ground. The rosette-like leaves develop later. Like most of the nightshade plants, this one also contains some alkaloids. Their concentration is so low, however, that the plant no longer is of particular importance in medicine.

The tubers of the green-winged orchis *(Orchis morio),* an orchid occurring in Central and South Europe and in Asia, growing in sunny meadows, on grassy slopes and in open woods (Ill. 148), were considered and unfortunately still are considered an aphrodisiac in some places. As a food or tonic for children and diseased people, the relatively big round

tubers were dug out, washed, sterilized by boiling water, and dried. Today, the tubers are still collected in Turkey and used to prepare dishes and drinks, which, taken especially in winter, are believed to "heal" colds.

As a consequence of wasteful exploitation and because of certain changes in biotope, the plant, formerly growing abundantly, has disappeared in many regions. Only by the strict observance of nature conservancy regulations can this plant be preserved for future generations.

Mandrake

The gist of the matter

After all these stories about magical and witches' plants, naturally the question arises whether these superstitious ideas have a real background. For a long time, scientists have been concerned with this question and have arrived at the conclusion that there actually is some truth in it. In 1925, the pharmacologist Hermann Fühner in a work on Solanaceae (nightshade plants), wrote the following about witches' ointments:

There is no doubt that the narcotic witches' ointment not only narcotized the victim but made him or her experience the whole beautiful dream of a flight through the air, of a banquet, of dance and love, in such an impressive form that he or she was fully convinced of the reality of the dream after waking up. In this way, the witches' ointment was a narcotic and luxury of the poor who could not afford more expensive pleasures... Noteworthy is the fact that this ointment was said to be capable of changing men into animals. German witches were believed to be transformed into cats, hares, owls, geese and other animals... Apart from Solanaceae, some of the witches' ointments also contained aconite (toxic constituent of the monkshood, *Aconitum napellus*). Just this addition of alkaloids exciting the sensitive nerve ends in the skin and then paralyzing them, could produce the autosuggestion of being transformed into an animal, of a coat of hair or feathers growing from the body, similar to hallucinations caused by skin irritations observed in cocainists today.

About fifteen years ago, the ethnologist Will Erich Peuckert and a friend subjected themselves to tests, using a witches' ointment the formula of which is still known. They applied it to their foreheads and armpits. After a short time, they fell asleep as if intoxicated. Later they woke up with severe headache and a dry mouth just as if they had been drunk. Peuckert himself reports the following:

We had wild dreams; at first horribly distorted faces danced before my eyes. Then, suddenly, I had the feeling as if I were flying many miles through the air. The flight was repeatedly interrupted by sudden falls. In the final phase I experienced the picture of an orgiastic feast with grotesque sensual excesses.

This experiment shows that witches' ointments prepared from plants produce a certain effect on the human body. Some other plants to which magical powers are attributed produce similar effects. Many of them excel in a peculiar scent that derives from the content of essential oils and other substances. In this connection we should like to remind the reader of garlic, sweet woodruff, caraway, thyme, and lovage which are used as spices. Other plants have played an important part in popular medicine and are still prescribed for various diseases.

It is certainly well known that in former times medicine and pharmacology were closely related to witchcraft and magic art. Moreover, magic plants include many toxic species where the connection with witchcraft is obvious. The belief that a small bunch of Saint-John's-wort protects from lightning stroke and the like is, however, a pure fiction.

Ingredients used in the preparation of drugs and for other medicinal purposes are contained in many plants. For ages, man has been trying to obtain help from such herbs. In spite of progress made in research and in the production of chemical drugs, pharmacies cannot dispense with plants, and the gathering of medicinal plants is still a paying occupation.

Many of the medicinal herbs used today were known in ancient times. The Greek physician Dioskurides, who lived in the first century, in his book *Comprehensive Pharmacology* gave descriptions of all medicinal plants known to him, including milfoil *(Achillea millefolium)*, eyebright *(Euphrasia officinalis)*, anise *(Pimpinella anisum)*, coltsfoot *(Tussilago farfara)*, lanceolate plantain *(Plantago lanceolata)*, blackberry, centaury, Saint-John's-wort, and lovage.

Fennel *(Foeniculum vulgare)* played an important part in the customs of the ancient Egyptians and Greeks, who appreciated it as a medicinal plant. Chamomile *(Matricaria chamomilla)* is known from the ancient Arab medicine (Ill. 145).

The use of arnica *(Arnica montana)*, foxglove *(Digitalis purpurea)*, and ergot *(Claviceps purpurea)* (Ill. 144) for medicinal purposes is more recent. The toxicity of the two latter plants has, however, been known for a long time, and both have been used for criminal purposes.

In ancient times people already knew that eating certain plants or parts of plants was harmful ans could lead to death under certain circumstances. Among the best-known poisonous plants of the Central European flora are foxglove, poison hemlock *(Conicum maculatum)* (Ill. 146), water hemlock *(Cicuta virosa)*, and meadow saffron *(Colchicum autumnale)* (Ill. 149), as well as the "witches' plants," deadly nightshade, henbane, thorn apple, and monkshood.

Hemlock was already ill-famed in antiquity. The Greek philosopher Socrates (469—399 B.C.), who focused his meditations on man and his activities, was sued for denying the gods and misleading the youth. As he did not revoke his teachings, he was sentenced to death. As was the custom of that time, he had to drink a cup of hemlock.

Hemlock is one of the umbel plants reaching a height of between 50 and 250 cm. In its lower portion, the grooved

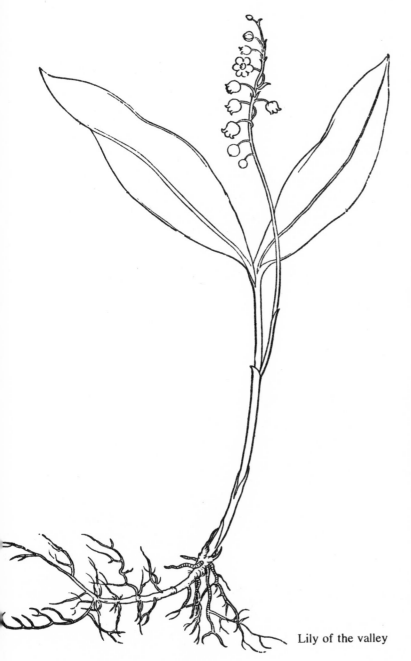

Lily of the valley

stalk usually is red-spotted and bears two- to fourfold pinnate leaves. The white flowers are arranged in large umbels and unfold from June to September. The plant mainly grows along hedges, walls, on waysides and fallows. It is distributed from Central Europe to Norway, Finland, to the Altai Range and the Lake Baikal region and southward to North Africa. The pungent smell of mouse urine is a characteristic feature of the plant. The toxic alkaloids contained in the plant paralyze the motor nerve ends so that death results from respiratory arrest, but the victim is fully conscious.

On the other hand, the poison of meadow saffron, the colchicine, acts on the central nervous system. As small an amount as 20 mg contained in about five seed kernels may kill a man in two to five hours. The mode of living of this plant is quite queer. From August to October it flowers in damp habitats, in meadows and grassland, from lowlands up to mountainous regions. During this time only the flowers are visible. They consist of six light violet petals grown together into a long tube at their lower ends; the stamens are adnate to the tubes and the ovary still rests in the earth. It is only in the following spring, when the flower stalk extends, raising the fruit including a large number of seeds above the surface of the ground, that the large, lanceolate, blunt, bright green leaves unfold. This habit has contributed to many superstitions. Foxglove with its strikingly red flowers growing on stalks that may reach a height of more than one meter, an embellishment of many mountain forests, has been known as a toxic plant for a long time, whereas it has been used as a medicinal plant only since the Middle Ages. However, a striking change took place. Doctors of the Middle Ages prescribed it as emetic or laxative. Today we know that these effects were due only to intoxication. In our time, the ingredients of foxglove occupy a prominent position in the treatment of heart diseases because they invigorate the heart muscle and influence the heart rhythm. As the content of active substances largely varies, depending on most different factors, wild plants are practically no longer used but foxglove is cultivated in fields.

Ergot is both a toxic plant and a drug; sometimes it can be found in rye ears and has the appearance of big, misshapen, dark rye kernels (Ill. 144). The plant is one of the fungi and develops in a complicated cycle. In this process, the ergot proper is a special permanent form. In former times, ergot was ground together with the grain; as a consequence, poisoning occurred and caused vomiting, chills, and blood flow disturbances in the extremities which sometimes became gangrenous and fell off. In old chronicles, this disease is called "Holy Fire." Today, ergot is used primarily in gynecology and obstetrics.

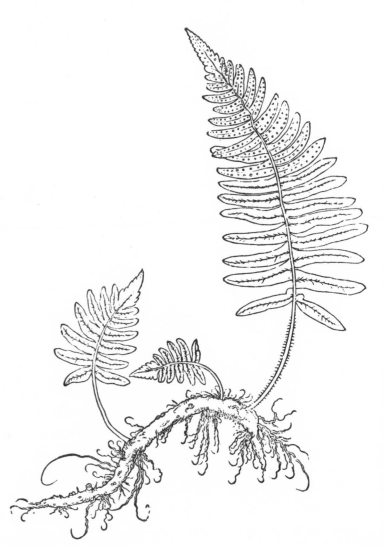

Blechnum

In former times, poisonous plants were used not only for criminal purposes or cultic actions, in some regions they were also of economic and military significance. Inhabitants of tropical forests used them for the preparation of arrow poison. In Asia and South America, some species of the genus *Strychnos* (*Nux vomica*, vomiting nut), which belongs to the logania family (Loganiaceae), were used to obtain arrow poison from their roots and barks (Ill. 150). In South America, *Chondrodendron* species of the moonseed family (Menispermaceae) were used for this purpose.

The use of poison arrows in South America has been known for a a long time, for in a report written in 1516 it states:

The Red Indians poison their arrow with a plant sap which leads to death.

The same report gives a description of the preparation of this poison:

As already mentioned, the Red Indians soak their arrows with a sap which they extract from various plants. But not everybody is allowed to prepare this mixture; only old women experienced in this field may do so. They are locked up with all essentials and have to remain awake for a period of two days when preparing the mixture.

Obviously, the preparation proper was very simple, but it was accompanied by quite a number of cultic and magical procedures. In essence, the chopped roots and barks were boiled, and the fluid extracted was boiled down to a syrup and mixed with the highly viscous sap of other plants to make the poison adhere to the arrowheads. The effects of arrow poisons, also known as curare, soon aroused the interest of scientists, who tried to use these substances for medical purposes. The famous French physiologist Claude Bernard (1813–1878) studied them and wrote in 1864:

When a mammal or a human being is poisoned with curare, his mental capacities, the sensitivity of his organs of sense, and his will power are not affected by the poison, but the organs of movement gradually fail to function. The most expressive abilities, above all voice and speech, disappear first, followed by the motility of the extremities, of the facial muscles, and of the thorax; finally, the respiratory movements, which are maintained up to the end, cease, as is the case in dying persons.

The Red Indians used curare not only for poisoning arrowheads but also for medical purposes, curing diseases of the stomach and epilepsy. At the beginning of the 19th century, medical men of Europe made successful experiments with curare in the treatment of tetanus. Yet the triumphant advance of this drug in medicine began only in the 1940s, when more exact information about the mode of action of this drug was available. By an interruption of the transmission of excitation from the nerves to the muscles in the motor end-plates, it leads to a relaxation of muscles, which with higher dosage, soon cease to work. This also defines the medical field of application. Curare is administered in the case of surgical interventions when everything depends not only on interrupting the sensation of pain of the patient but also on immobilizing and relaxing the muscles. Normally, quite small doses will be sufficient for these purposes, and, when handling this poison, medical men have to take every precaution. Today, curare is frequently replaced by synthetically obtained substances of a similar chemical composition.

Ginseng between superstition and reality

The root of ginseng *(Panax shin-seng)* is called a magical drug, an inexhaustible elixir of life, a herb of eternal youth. According to legend, the Chinese thinker Lao-Tse had discovered the drug for long life, the Jen-Shen, already 2,300 years ago. Another legend tells that the omnipotent mountain ghost had sent a rescuer to distressed mankind—a boy in the shape of a human-like turnip, the ginseng root. This root has enjoyed highest repute in Korea, China, and Japan for about 4,000 years. In the Chinese 52-volume work *Pent-ts'ao kang-mu* (Classification of Roots and Herbs), the experiences gathered in the course of thousands of years were compiled for the first time; the work was published in 1597.

According to this work, ginseng is a remedy for the lung, invigorates the spleen, cools the fire, and automatically opens the heart; it enriches the knowledge, strengthens mental capacities, soothes shock, removes high temperatures and diarrhea, causes the blood to circulate in the blood vessels, removes constipation, and causes accumulated slime to flow off. Ingested in a boiled and pappy condition, ginseng can renew the vital power even if it has almost disappeared. It stimulates the "five intestinal organs," rejuvenates the body, and extends life.

The Dutch brought the ginseng root to Europe in 1610. It became rapidly known under the name of pentao and was used as an aphrodisiac at the court of Louis XIV.

Ginseng grows wild in the Ussuri region, in Manchuria, and in Korea. Today, it is also cultivated in some regions of the Soviet Union, in East Asia, and in North America. As a shadeloving plant, it mainly grows on the northern slopes of deep mountain forests. As it has been collected for a very long time, it has almost been eradicated in nature and, today, is found only in places that are almost inaccessible.

Ginseng is a shrub, 30 to 60 cm high, having a branched root resembling a carrot that is yellowish-white in color. It sometimes has the appearance of a mandrake. The stalk bears the five-foliate ovolate leaves; it has umbels of small greenish flowers succeeded by scarlet berries in autumn.

Some time ago, ginseng was examined for its active substances, especially by Soviet scientists. The results obtained confirmed the thousands-of-years-old ideas of its great curative powers so that it is reckoned among the medicinal plants today. It is mainly used to cure neurasthenia and fever.

Ginseng plant *(Panax shin-seng)*

Not only botanists, pharmacists, medical men, and etnologists but also Interpol and the police of many countries are concerned with quite a number of plants associated with superstitious ideas. The plants involved are those that contain narcotics.

It has been known for a very long time that certain plants contain substances producing a narcotic effect. The ancient Greek historian Herodotus reported the following:

At the edge of a small forest stands a low structure, half cottage and half tent, covered with felt on all sides. People seem to gather here. They pass the entrance in a stooped attitude, the priests being the last to enter. Soon white smoke is emitted from the cracks, and cries of joy can be heard. The ceremony is in full progress.

Let us enter. In the center of the room is a hearth covered with large stone plates. It is used as an altar on which hemp kernels are spread as soon as the plates are red-hot. The rising narcotic vapors quickly intoxicate the faithful. They show serenity and are seized by an ecstasy of joy: a proof of the presence of the gods and their favor. They had come down to their brave Scythians.

The plant used by the Scythians more than two thousand years ago still plays an important role as a narcotic. It is Indian hemp *(Cannabis sativa* or *indica)*, which, as a drug, is called hashish or marihuana, to mention but the most frequently used names (Ill. 151). Bhang, known from Oriental fairy tales, is nothing else but hashish. As a fiber plant, hemp is one of the oldest cultivated plants and, in our time, is grown in almost all countries with a warm climate. It is a dioecious plant and reaches a height of up to 4 m. Fibers are obtained from hemp by various methods and are manufactured into coarse fabric, ropes, and the like. The narcotic and intoxicant substances are contained in a few species in which they are found in small glands between epidermis and cuticle in female plants. These glands are partly stripped off, and the hashish is collected in this way. More frequently the dried leaves are smoked alone or mixed with tobacco. Formerly, hashish was also eaten, for example, added to jam.

After ingestion hashish produced hallucinations of a "paradise" in the addict, but some writers claim that, like all such drugs, it leads to habituation and finally brings with it severe physical and mental damage which results in a general deterioration.

Opium, also known from ancient times, is a drug derived from the dried milky juice of the opium poppy *(Papaver somniferum)* obtained from incisions made in the unripe plant capsules. It was used by Assyrians, Egyptians, and Greeks, and many physicians prescribe it as a narcotizing drink and for the soothing of pain. These effects of opium are the major reasons for its wide distribution. Morphium and heroin, more dangerous and effective narcotics, are made of opium. For a long time, opium has been used for medical purposes only; its misuse probably began toward the end of the 17th century in the form of opium smoking. Mode of action and results are similar to those of hashish.

For several thousands of years, the Mexican Indians have known several vegetable intoxicants among which peyotl or peyote ranks first and foremost (Ill. 152). This small dark blue-green cactus, also known as mescal, has a flat spherical shape and a turnip-like root. It is native to the region from Central Mexico to South Texas. Depending on the age of the plant, it has five to thirteen flat ribs divided into tubercles or buttons by transverse furrows. Ambassadors of the Ancient Spanish Indian Council, clergymen and explorers, reporters and writers of adventure stories, chemists and medical men, state authorities and smugglers have been concerned with this legendary cactus, so that a comprehensive literature on this subject exists in which truth and fantasy form a strange mixture.

The ancient Mexican peoples considered peyotl *(Lophophora williamsii)* as a god. At that time intoxication was something extraordinary for the Indians. This explains the central position occupied by the cactus yielding an intoxicant in the ancient Mexican heaven of cacti gods.

The famous cacti specialist Curt Backeberg comments on the mode of action of mescal, the substance of peyotl, in his book *Stachlige Wildnis*:

When eating dried pieces of its turnip-like body, quite odd hallucinations occur. Initially, medicine men may have administered them, unsuspecting, as a remedy against diseases like any other herb of which healing properties were expected. When they saw, however, what the drug's effect on man was, they were convinced that this could be produced only by a god. He bestowed on the mescal eater unusual endurance, he produced by magic gorgeous color visions and strange melodies in the environment of the mescal eater who was fully conscious, the god made him bubble over with joy and experience overwhelming apparitions. Peyotl certainly was the god of the soil because he crept so deep into it that he was almost invisible. One imparted the shape of a stag to the omnipresent deity, hunted him with a special bow without injuring him, that is to say, one darted arrows near him into the ground toward the four cardinal points as a symbol of the embedding of the seeds into the fertile soil, and one composed songs in his honor. When the Huitcholes migrated from the area that today is the state of San Luis Potosí to Nayarit, every year a group of them, the Peyotleros, returned to the old home in order to take the hundredfold god to the new dwelling place with certain ceremonies, him who is not so great as father sun but sitting at his right side. This custom has been maintained to this day. A number of those Indians still go to the mountains where they stay for many weeks toward the end of the year; when they again appear in their villages, about the beginning of December, one evening the great peyotl feast is held with music and dance until men and women withdraw, disappearing in the gloom of the huts to devote themselves to mescaline intoxication.

So great is the power of Peyotl that he overcame the resistance of the Christian priests and even found a way into the new doctrine. He has proved to be that one of the ancient cacti gods who outlived all others because he is still capable of spellbinding many people.

A particular feature of this cactus, which has no prickles, is its content of several alkaloids, of which mascaline and anhalo-nine are the most important. They lead to euphoric states of intoxication associated with hallucinations, color visions, and delirium. Larger doses produce symptoms of paralysis and can be lethal.

A large number of reports by writers and physicians on the effects of peyotl are known to us. One of the most informative was given ba the English physician and writer H. Ellies:

I saw jewelry, singly and in packed groups, and wonderful carpets. Now they gleamed in a thousand fires, now smouldered with a dark and glorious shimmer. Then they transformed themselves, before my eyes, to flowers, to butterflies, and to glittering wings. With every second that passed I was regaled with new forms. Now darkly gleaming colors, alive with movement, and a specially magnificent one seemed for an instant to approach me, then they gleamed with fire, then they shimmered. Most appeared restrained combinations of color, with shining dots, like jewels.

Besides its use as a cardiac, one substance contained in peyotl, the mescaline, has gained in importance for other reasons in recent years. The visions produced after ingestion of mescaline in many respects resemble the visions associated with certain mental disorders. That is why, today, mescaline and substances producing a similar effect are used for the study and treatment of mental disorders.

With this, the circle encompassing witches' herbs, poisonous plants, and drugs is closed again.

141 Saint-John's-wort *(Hypericum perforatum)* was said to protect against lightning stroke.
142 The alkaloid of monkshood *(Aconitum napellus)* was a constituent of many witches' salves.

143 In antiquity magic powers were ascribed to the root
of mandrake *(Mandragora vernalis)*.
144 Ergot *(Claviceps purpurea)* is both poisonous and medicinal.

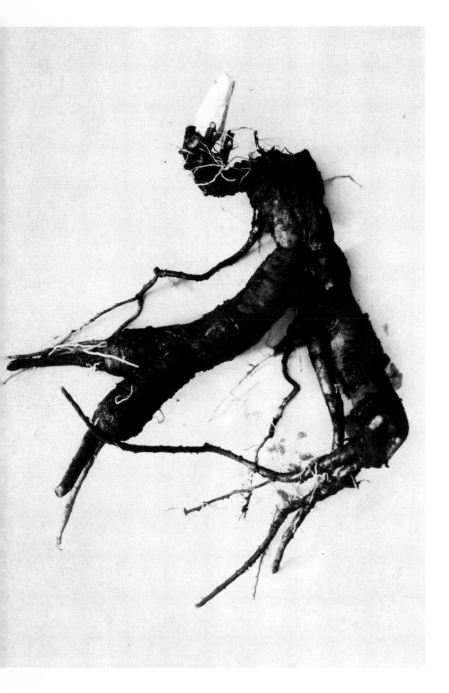

145 Chamomile *(Matricaria chamomilla)* still is used as a medicinal plant.
146 The spotted hemlock *(Conium maculatum)* is a well-known poisonous plant.

147 The peony *(Paeonia officinalis)* is not only the most showy item of many gardens but was used as a medicinal plant and, formerly, as a magic plant.

148 For a long time the tubers of the green-winged orchis *(Orchis morio)* were considered aphrodisiac.

149 Special forces were attributed to meadow saffron *(Colchicum autumnale)* because of its special development.

150 Several varieties of nux vomica *(Strychnos)* are used as arrow poison.

151 Indian hemp *(Cannabis sativa)*
left ♂, right ♀

The first traces of life

In the Carboniferous forest

The fossil book
of nature

Seed formation

A strange discovery

The great change

In the preceding sections of this book, we have been given a general idea of the immense variety of forms of plants existing today. When considering the different habits, and the multifarious flowers and fruits of trees and shrubs, we should bear in mind that they are the result of a development of millions of years and that all higher forms arose from lower ones.

The study of the plant kingdom of former periods of the earth's history is an interesting and complicated field, which is known as paleobotany, a special branch of the science of botany. The most important material subjects of these studies are the fossil remains of past geological ages. They are available in the form of dendrolites—carbonized, mineralized, and petrified remains of plants—and in the form of inclusions in amber—spores or pollen grains.

In ancient Greece, Xenophanes and Theophrastus already 2,500 years ago described such fossil plant remains. They were also known to the painter Leonardo da Vinci (1452–1519) and to the founder of scientific mineralogy and the science of mining, Georgius Agricola (1494–1555), as some publications and illustrations show. Nevertheless, the first comprehensive paleobotanic work was published no earlier than 1709. Its author, Johann Jakob Scheuchzer, called it *Herbarium dilivianum* (Herbarium of the Deluge), because at that time scientists held the view that fossils were the remains of plants and animals that had lived before the Deluge and perished during this flood.

Toward the end of the 18th century, the scientific study of the fossil material began, a study that is closely associated with the name of Ernst Friedrich von Schlotheim (1764–1832), who published his book *Beschreibung merkwürdiger Kräuter-Abdrücke und Pflanzenversteinerungen — Ein Beitrag zur Flora der Vorwelt* in 1804 and the standard work, *Die Petrefaktenkunde auf ihrem jetzigen Standpunkte,* in 1820.

In the second half of the 19th century, the paleobotanic research made considerable progress when the theory of evolution established by Charles Darwin (1809–1882) was gradually accepted. Today, paleobotany gives a clear survey of the development of the plant kingdom and of the flora of the individual periods of the history of the earth.

Little is known of the oldest living beings and of the oldest plants. They were delicate and unstable so that they could not leave traces in rock. Today, however, one knows bacteria-like fossils, however, which were found in stones of the Figtree Series in South Africa that are 3.1 thousand million years old. The Bulawayo limestone in Southern Rhodesia and the Soudan-Iron Formation in Minnesota, both of which are 2.7 thousand million years old, contain biological material. The algae-like deposits of Lake Superior (Ontario) surely are plant remains; they are estimated to be 1.9 thousand million years old. All of these forms can give us only a rough survey of the time of occurrence of certain more comprehensive groups of plants, because their state of preservation does not allow more detailed examination.

This, however, does not apply to the first land plants, which appeared about 400 million years ago, and not only in one place on the globe but in several places simultaneously. Their remains have been found in Europe, North America, Australia, and a few other regions. They are so-called psilophytes, dichotomously branched, small plants that were leafless or only covered with scales, and that multiplied by spores. Probably they grew at the edges of swamps or bogs and thus were in close connection with water. They had no roots. These plants lived in geologic periods that are termed Ordovician, Gotlandian, and Lower Devonian.

The era of the cryptogamous plants extended from the Upper Devonian period to the Lower Permian, the Rothliegende period. A large number of plants of this era were preserved, so that we have a rather clear idea of the vegetation of that time. The wealth of plants existing in the Upper Carboniferous period is particularly obvious, because the coals from the Carboniferous Period are, like all other pit coals, the remains of fossil plants. That is why we speak of Carboniferous forests, which differ from our present forests in a few features because the former comprised mainly club moss-like, horsetail-like, and fern-like plants, whereas the seed plants were represented by only a few gymnosperms, and angiosperms were still entirely missing.

Now let us consider some representatives of the Carboniferous flora and compare them with their relatives living today.

In the Carboniferous period, plants of the club moss family best known to us were trees of the genera *Lepidodendron* and *Sigillaria*. They attained a height of 20 to 30 m and a stem diameter of up to 2 m. The peculiar structure of the tree was due to the scars left by shed leaves, which attained a length of up to 1 m. The trees were anchored to the soil by a flat rootstock, a fact that indicates a swampy habitat. The spores formed cones which, in the lepidodendrons, were borne on the outer tips of twigs or on particularly short branches and, in the sigillaria, directly on the stem. This also is a form of cauliflory that is still found today in trees living in tropical forests (cf.p. 11).

The club mosses occurring today have practically undergone little change since that time. They are, however, considerably smaller and play an entirely subordinate role in vegetation as inconspicuous, evergreen, herb-like plants with dichotomously branching sprouts and small scaly or needle-shaped leaves. Mostly spores are formed in special capsules or sporangia borne by sporophylls.

The sporophylls of *Huperzia selago*, a club moss species, have the appearance of leaf buds (Ill. 153). This species of club mosses is the most primitive of those living today, seen from the aspect of the history of evolution. It grows in damp and shady woods, occurs very sparsely in the temperate zones of the northern and southern hemispheres, and is missing in many places, especially on level land. In mountainous regions it grows up to an altitude of 3,400 m and also occurs in the Arctic. The upright stalks attain a height between 5 and 30 cm. Because of its rareness, *Huperzia selago* is protected by nature conservancy regulations in many countries.

In the Carboniferous period, horsetail-like plants were practically as abundant as club mosses. In this connection, Calamites played an important part; they attained a height of 20 to 30 m and a stem diameter of 1 m. The stem rose from widely spreading rhizomes and exhibited a remarkable growth in thickness. The leaves were seated in whorls borne by side branches likewise arranged in whorls. The sporophylls formed ears on the upper nodes; hence, this also was a form of cauliflory. As the stems of the Calamites were hollow, so-called pith stone cores, stony fillings of the inner pith cavities, many of them have been preserved. Fossils of small herb-like and liana-like horsetail plants from that time do exist, a fact that allows us to draw conclusions about the great variety of forms of this group during the Carboniferous period.

Only about 30 species of this plant group, which was very comprehensive in the days of old, are still in existence; all of them belong to the genus horsetail *(Equisetum)*, whose general structure largely resembles that of the extinct species. The extant species are perennial herbs with creeping rhizomes, though considerably smaller than their extinct relatives. A few species are green only in summer and have tuberous rhizomes. The sprouts are clearly arranged in nodes and internodes. The usually scaly leaves form whorls, are connate at their bases, and enclose the stalk in the form of a sheath. In contrast to the Calamites, the cone-shaped sporophylls are arranged at the end of the stalks. An interesting phenomenon is the rapid growth of the sprouts in spring. They are already formed in the preceding vegetation period and in spring have only to stretch.

The largest recent horsetails still are almost arborescent like their ancestors. They attain a height of up to 10 m and grow in tropical South America *(Equisetum giganteum)*. The giant horsetail of Europe and North America *(Equisetum tel-*

mateja) is the most impressive form (Ill. 154). The vegetative sprouts attain a height of 2 m but are only 10 to 15 mm thick. The plant grows in moist, fresh, and calcareous soils.

The first ferns known appeared in the Devonian period, and during the Carboniferous period they showed a large variety in forms. Many of these forms died out at the end of the Paleozoic era, in the Permian period. In the following periods of the history of the earth, new groups developed again. Today, about 12,000 species of fern are distributed all over the earth but occur mainly in tropical zones, where arborescent forms also grow (Ill. 157).

Besides club moss and horsetail plants, arborescent ferns determined the appearance of the Carboniferous forests. Whereas club moss plants had bark stems and horsetail plants tubular stems, ferns developed root-type stems. This ensured a better water supply to the plants. Simultaneously, ferns were the first plants to develop pinnate and large-area leaves, which made them superior to other plants with regard to the uptake of carbon dioxide and air, as well as utilization of light.

Within the ferns, panicle ferns form a relatively isolated group. They existed in the Upper Carboniferous period, reached the climax of their development in the Jurassic period, and are represented today by about 20 species distributed all over the globe. Among them, there is the most imposing fern of Europe, the royal fern *(Osmunda regalis)*, which forms large bushes with its up-to-2-m-high fronds (Ill. 155). It grows in trenches, shrubs, and forests, especially in boggy soils with a low lime content or in soils with stagnant water. It occurs in Western Europe down to North Africa, in the Atlantic Coast areas of North America and central South America, in South Africa, southwestern India, and East Asia. Whereas it is missing in the mountainous regions of Central Europe, in North Africa and North America, it exists at altitudes of up to 2,000 m. In some places it is very rare, and in a few countries it is protected by nature conservancy regulations.

The fronds are very decorative. They are arranged in the form of a funnel, with double-pinnate leaflets that are green in summer. The inner fronds of each funnel bear fertile leaf sections toward the top and pure sporophylls at the top, which are located in a much-branched panicle. Originally they are green, becoming brown later. Because of their loose condition and their long period of durability, their rhizomes are used in flower gardening as a substrate for the cultivation of epiphytes, especially tropical and subtropical orchids. Because of ruthless exploitation, many of the former habitats became completely devoid of them.

Reconstructed trees of the genus *Sigillaria*

In the Upper Carboniferous period, the first seed plants appeared. This change in the development of the plant kingdom became necessary because of climatic changes. As a consequence of more pronounced climatic differences, greater seasonal contrast, as well as the drying up of large lowlands, the mode of multiplication of plants had to change basically, and new forms of adaptations had to be developed. The plants that had existed up to that time were cryptogamous, with the production of a new generation not being possible without water because the male germs moved through water to the female ones. In the new seed plants, however, the process of fertilization took place in the plant itself so that they were independent of water outside the plant. With the seeds formed, the plants also obtained a new means of distribution.

This new method of multiplication was indicative of the transition to a new era of plant development (mesophytic) which extended from the Upper Permian period to the Cretaceous period. During that time, the predominant group was the gymnosperms, which included, among others, the cycads and conifers, and the ginkgos. A very interesting group of plants, which also are among the gymnosperms, are the seed ferns. They have fern-like leaves but are true seed plants according to their ligneous structure and their mode of multiplication. These plants died out millions of years ago. Probably they appeared for the first time in the Upper Devonian period and their last representatives grew in the Jurassic period. Particularly large numbers occurred in the Carboniferous period and in the Rothliegende period (Ill. 156).

Externally, the seed ferns probably were similar to the members of a plant group that is still found today but look rather primeval—the cycads, of which 10 genera comprising about 100 species are still in existence. They reached the climax of their development during the Mesozoic era; from that period a large number of fossils have been found. The appearance of the cycads resembles a few species of palms (Ill. 159, 160). The sturdy, usually nonbranched lignifying stem attains a height between 2 and 15 m and is covered with the scars of shed leaves. At the top, it bears a tuft of large frond-like, spirally arranged and single- or double-pinnate leaves. The plants are dioecious; they form cones of different sizes, which are spirally arranged at the end of the stem axis (Ill. 161). Pollination is effected by the wind, and occasionally by beetles. A peculiar feature of cycads is the roots. A taproot reaching deep into the soil puts forth normal side roots and forked side roots that lie close to the soil surface, not growing downward but toward the soil surface and partly extending into the air. In these roots, algae live which bind atmospheric nitrogen, thus ensuring an additional supply for the plant.

Like cycads, ginkgos form an isolated group among the gymnosperms whose main developmental time was from the Jurassic period to the Lower Cretaceous period.

Today, the only representative of this group of plants, which once was very rich in forms and was distributed all over the world, is the maidenhair tree *(Ginkgo biloba)*. This tree is now in existence only in a small area in East Asia (Ill. 158). In the Tien-Mu-Shan mountains in Southeast Asia, the last wild-living representatives of this plant, which is of particular value seen from the angle of the history of evolution, are found. In parks and gardens all over the world it is again to be found today—since 1730 in Europe and since 1784 in America.

Probably the maidenhair tree is one of the oldest still living species of plants. That is why Darwin already called it a "living fossil." The heavily branched tree, green in summer, attains a height of 30 m, bearing long and short shoots. The short twigs of the widespread branches bear the typical bilobular leaves with distinctly fork-like and fan-like extending nerves. In autumn, the leaves become golden yellow in color. The dioecious flowers develop in the axils of leaves. Pollination and fertilization are still rather primitive.

In front of the house in which Goethe lived in Jena a maidenhair tree was growing. He considered the leaf of the maidenhair tree a symbol of inner conflict in man. In his *West-östlicher Diwan*, he wrote the following poem:

Leaf from this tree, that from the East
comes and takes root in my garden,
gives food for thought,
and edifies the knowledgeable man.

158

Is it one living thing,
Divided in itself?
Are they two, each one the other's choice,
So that we see the two as one?

To unriddle such questions
Is not hard for me.
Can you not feel, in my song,
That I am one, yet duple?

The most important gymnosperms are the conifers which first occurred in the Upper Carboniferous period and still play an important role in the flora and vegetation of our time. The first conifers were small trees that grew outside the bogs, but soon larger forms developed; in the environs of Karl-Marx-Stadt (GDR), fossil stems that attained a height of up to 20 m have been left from the Rothliegende period. As a unique natural monument, some of the finest and largest stems of them can still be admired in front of the Karl-Marx-Stadt Natural Science Museum as so-called petrified wood (Ill. 163). As these stems are true petrifications, details of their structure, even including the cells, can be seen in thin-ground sections. Besides stems of conifers, petrified remains of stems of ferns, seed ferns and horsetail plants were found near Karl-Marx-Stadt. In the Lower Triassic formation of Arizona petrified woods were discovered, too. The biggest fragments found have a diameter of 1 m and a length of 2.40 m. Their age is estimated to be about 180 million years. The place of discovery was declared a nature preservation zone. Further "petrified woods" exist in the vicinity of Cairo and in Patagonia. They are ascribed to the Tertiary era and have an age of about 30 million years.

Other finds of leaves and cones permit us to form a rough idea of these trees today. Probably, they resembled externally—and only externally—the indoor plant araucaria *(Araucaria).*

Since the Jurassic period, a few other groups of conifers have been occurring in greater numbers, including the araucarias, the bald cypresses, the mammoth trees, and the umbrella pines. Like the maidenhair tree, the umbrella pine *(Sciadopitys ver-*

ticillata) is considered to be one of the oldest still living species of plants (Ill. 162). In the Middle Tertiary period, the plant occurred in such great numbers that its typical double needles formed a lignite layer of its own in several places—the so-called grass coal. Deposits of this grass coal are found, for example in Lusatia and in the Rhineland. Today, this species occurs only in southern Japan, in temperate regions at altitudes between 300 and 1,500 m. This huge tree, having a narrow pyramid-like crown, attains a height of up to 40 m. The tips of the short annual shoots bear the characteristic umbrella-like, bright green double needles, which show a deep groove in their centers. Because the wood of these trees is widely used as timber for the construction of houses in Japan, the total number of trees has been considerably reduced. Although the umbrella pine, a beautiful conifer, is cultivated in many parks, there is the great danger that it will be eradicated as a wild growing tree.

Reconstructed tree of the genus *Lepidodendron*

A unique event in the history of botany was the discovery of the dawn redwood *(Metasequoia glyptostroboides),* which was found as a fossil about 30 years ago, and as a living plant a short time later (Ill. 164). It was in 1941 when the Japanese botanist Miki, studying fossil conifer twigs and cones from the Neocene in Honshu in Japan, discovered that a few species had opposite leaves and long-stalked cones, but not alternate leaves and short-stalked cones like the other conifers. This induced him to establish a new genus which he called *Metasequoia.* Since that time, fossil material of the genus *Metasequoia* has also been reported from other places and strata of discovery, so that we know today that it grew in the Upper Cretaceous period and occurred not only in East Asia but also in North America, northern Siberia, and Greenland. In the winter of 1941–42, T. Kan of Nangking University made a journey through the Chinese provinces of West Hupeh and East Szechwan. Near the village Ma-tao-chi he found a needle-shedding tree, unknown up to that time, which was called *Shui-hsa* (water larch) by the local people. As no material was collected, the plant could not be defined exactly. In the summer of 1944, T. Wang of the Central Bureau for Forestry Research in Nangking traveled through the same region and collected twigs and cones of this tree species, which at first was considered a new species of the genus *Glyptostrobus*, the water pine, which also grows in East Asia. In 1946, another expedition traveled into the almost inaccessible territory and counted a total of only 25 species of this new and interesting plant. Part of the comprehensive botanic material collected by this expedition was given to the director of the Arnold Arboretum of Harvard University (USA) who recognized the great importance of the finds and suggested and organized further excursions for the gathering of seeds. In this way, it was found that some 1,000 trees of different ages grew on an area of about 800 square km on slopes, in crevices, near brooks, and in paddy fields at an altitude between 700 and 1,500 m. Today we know that the total number is considerably larger. During the excursions, large amounts of seeds capable of germinating were collected and sent to botanical gardens and arboretums in many countries. In 1948, the botanists H. H. Hu and W. C. Cheng described the plant as *Metasequoia glyptostroboides.*

The dawn redwood attains a height of up to 50 m and a diameter of the stem of more than 2 m. The plant grows very quickly, and in its structure resembles the bald cypress *(Taxodium)* and the mammoth trees *(Sequoia, Sequoiadendron)* of North America. The needles, which are green in summer, are shed together with the short shoots in autumn. The long-stalked female cones are spherical or cylindrical.

Tests have shown that the dawn redwood is distinguished by an extraordinarily voluminous and rapid growth. Thus, 25-year-old trees have a stem diameter between 30 and 40 cm and attain a height of 18 m. Compared to this performance, this production of wood is achieved by a Douglas fir *(Pseudotsuga menziesii)* in 30 years, by a spruce *(Picea abies)* in 50 years, by a red beech *(Fagus sylvatica)* in 60 years, and by a fir *(Abies alba)* in 70 years. Besides this considerable growth, the dawn redwood is also distinguished by a great industrial hardness, low inflammability, and other valuable properties, so that many countries have conducted tests for cultivation in forestry for some time.

The umbrella pine *(Sciadopitys)* and the dawn redwood *(Metasequoia)* had a considerably larger area of distribution in the Tertiary Era than today, as already stated. These were no isolated cases. For example, during the Tertiary period, numerous plants occurred in Central Europe which today are found in North America or East Asia only. These include the maidenhair tree *(Ginkgo)*, the bald cypress *(Taxodium)*, the mammoth trees *(Sequoia, Sequoiadendron)*, the arbor vitae *(Thuja)*, the water pine *(Glyptostrobus)*, the Weymouth pine *(Pinus strobus)*, and quite a number of conifera. This phenomenon is not restricted to conifers but is also found in angiosperms as a result of the different effects of the glacial period.

The large mountain ranges in North America and East Asia run from North to South. In this way, they did not form an obstacle during glaciation as was the case with the European mountains. Because of this arrangement, the plants could move toward the South and later return to their original area. In Europe the greater part of the Tertiary flora was destroyed.

The great change

One of the greatest changes in the history of development of the plant kingdom took place toward the end of the Lower Cretaceous period, about 135 million years ago. Gymnosperms receded, and angiosperms became the dominant plant group. This was the beginning of a new era in the plant kingdom coinciding with the beginning of the Cenozoic era. Within the short period of *only* 60 million years starting with the Tertiary period, a vast number of completely new families and genera with hundreds of thousands of species developed. Of the gymnosperms, only a few hundred species have survived.

In that time, that great change took place when the present mountains and the present deep-sea basins came into being. Within this period of millions of years, the face of the globe—continents and seas, relief and climatic zonation—was changed, with differences between climates and seasons of the year becoming more pronounced. It was toward its end that the diluvial glacial period began, which ended only a few thousand years ago.

Today, angiosperms determine the characteristic picture of vegetation in almost all zones. Among them there are trees, shrubs, and herbs, including almost all cultivated plants. In comparison with all other groups of plants, angiosperms are best adapted to the environmental conditions of the present time. Sexual reproduction is the most essential progress. Their ovules are always enclosed in an ovary which is formed by carpels and which alone, or together with other floral parts, forms the fruit after fertilization. In most cases, at least in the more primitive forms, the flowers are hermaphrodites, that is to say, they possess ovules and stamens. Owing to this and other developments, they are capable of a considerable acceleration of sexual reproduction. This, however, is not the only advantage of angiosperms. They are superior to other plants with respect to their system of water conduction, to the structure of their leaves, and in other aspects.

The origin of the angiosperms is not fully understood as no clear transitional forms from the gymnosperms have been found. Most scientists, however, believe that the ancestors of the angiosperms were plants that resembled the seed ferns.

Among the angiosperms living today, magnolias show the greatest number of original features (Ill. 165, 166), One of the most important of these features is the spiral structure of their flower, in which the number of the individual parts usually is large and varying. The stamens and carpels are arranged in the form of a spiral about the stretched axis, whereas in the higher plants they are arranged in circles in a small but constant number, with the axis being shortened. In the magnolia, the strongly built perianth is not yet divided into calyx and corolla (Ill. 167).

Like many other plants, magnolias grew in the whole northern hemisphere during the Cretaceous and Tertiary periods. At present, areas of distribution are found only in East Asia and North America. An interesting phenomenon is the fact that the East Asian magnolias flower in spring before the development of leaves, whereas the North American species flower in summer after the formation of foliage.

153 The most primitive club moss
still living today is *Huperzia selago*.
154 The vegetative shoots
of the giant horsetail *(Equisetum telmateja)*
grow up to 2 m high.

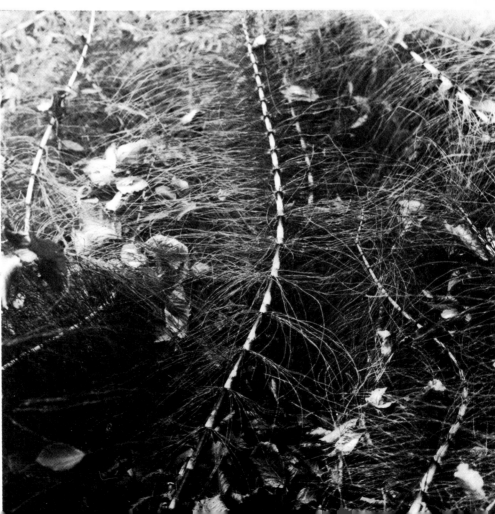

155 Royal fern *(Osmunda regalis)*
156 Impression of a fossil seed fern
(Neuropteris)

157 Tree ferns still grow in the tropical rain forests.

158 The maidenhair tree *(Ginkgo biloba)* probably is the oldest type of plant still occurring today.

159 The palm ferns include *Encephalartos transvenosus* of South Africa.
160 The palm fern *Microcycas calocoma* occurs only in a few specimens in the mountains of western Cuba.

161 Female inflorescence of *Encephalartos caffer*
162 Part of a shoot of *Sciadophitys verticillata,*
the umbrella pine
163 The "petrified wood" of Karl-Marx-Stadt is an almost unique
natural monument.

165–166 The flower and fruit of magnolias *(Magnolia grandiflora)* characterize them as primitive representatives of angiosperms.

About giant and millenarian plants and other curiosities

In the preceding chapters of this book, we have become acquainted with many a strange plant, such as the most different vegetational forms, bizarre flowers, specialists in the plant kingdom, magical plants, and fossils. But all these forms were nothing unusual, as they almost always were within the conceptions we commonly form of plants.

There are, however, also plants that are quite unusual, some with respect to their size, some with respect to their age, and some because of their strange shapes.

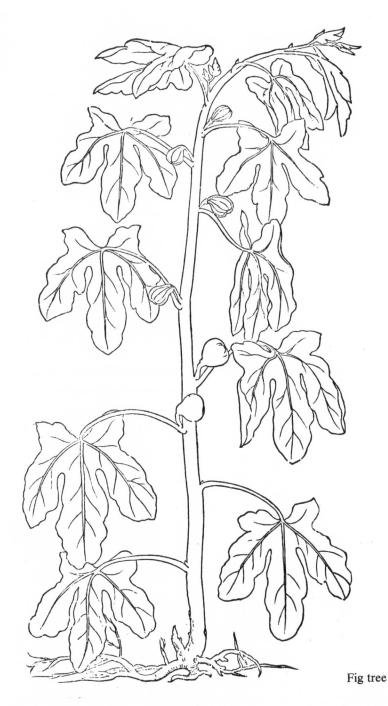

Fig tree

When speaking of large plants, we first think of trees, and this is quite correct, because among them there are the largest forms of plants. This, however, is not always true of European trees, although among the trees of Europe are some giant tree veterans, too. For example, the silver fir *(Abies alba)* reaches a maximum height of 75 m, and the spruce *(Picea abies)*, 60 m. The pedunculate oak *(Quercus robur)* and the common pine *(Pinus sylvestris)* can attain a height of 50 m. These figures, however, are maximum values that are reached in individual cases only. Usually the trees remain considerably smaller.

Among the most impressive and famous trees of the earth are the mammoth trees in North America which attain a height of more than 100 m and a stem diameter of 12 to 15 m. A distinction is made between two species of mammoth trees. The mammoth tree *(Sequoiadendron giganteum)* grows in the Sierra Nevada in Central California in more or less large areas at altitudes from 1,400 to 2,600 m above sea level. The total area in which this tree grows is estimated at 7,000 hectares. The whole area is a nature reserve (Ill. 169). A few particularly conspicuous trees were given special names (Ill. 168). The thickest mammoth tree living at present, "General Grant," has a height of 89 m and a stem diameter immediately above the ground of 12,5 m. The tallest of its kind is the "House Tree," which has a height of 92 m and a stem diameter of 8 m. Each of them is believed to be about 3,500 years old. The thickest mammoth tree ever measured, which no longer exists, is said to have had a stem circumference of 46 m. This corresponds to a diameter of 14.45 m. The tallest evergreen mammoth tree ever measured, "Father of the Forest," which unfortunately no longer exists, was 125 m high. This shows that this real giant among trees had attained a height which was double that of most of the tallest European trees. Its age was estimated at 3,200 to 3,800 years.

The evergreen sequoia *(Sequoia sempervirens)* grows in the coastal region of the Monterey Bay up to the northern boundary of California on mountain slopes in a narrow tract of land at altitudes of about 1,000 m above sea level. There it forms dense pure stocks that cover an area of 810,000 hectares. An area of 27,500 hectares is at present protected by nature conserv-

ancy regulations. This species exceeds even the mammoth tree in size. Among the trees still living heights of 112 m and 107 m have been measured.

The growth and the increase in height of these giant trees are of particular interest. Within 30 years a mammoth tree can attain a height of 24 m and a diameter of 40 cm.

In 1769, the first report of these giant trees was written by Father Juan Crespi; because of the color of their wood he called them redwood. Scientific descriptions of these plants, however, followed much later—in 1824 for the evergreen mammoth tree by the English conifer specialist A. B. Bourke, and in 1853, for the mammoth tree by the well-known English botanist John Lindley. The name sequoia, from which later *Sequoiadendron* was derived, was given to these trees after the name of the Cherokee chief Sequo-Ya, who developed the first Red Indian alphabet and introduced the script to the Red Indians.

Protection of the unique trees was initiated only after 1900. Both the then President, Theodore Roosevelt, and the National Geographic Society of California advocated the establishment of national and state parks as well as national forests in order to preserve these plants for future generations. Today, the evergreen sequoias grow in twenty coastal Redwood State Parks, of which the Humboldt State Redwood Park is the largest, covering 9,500 hectares. The national parks in the Sierra Nevada serve for the protection of the mammoth tree. The most important are the Yosemite National Park, the Sequoia National Park, and the Valaveras National Park.

Mammoth trees, however, are not the only giant trees of the earth. The Australian eucalyptus trees can compete with them in height. Some scientists even assume that a few *Eucalyptus* species grow even higher. But concrete statements have not been made so far. In the first chapter of this book we dealt with the occurrence and importance of this genus. We would just like to add that not all species of the genus *Eucalyptus* are tall trees. Many of them also occur in the form of shrubs. Tall trees are developed especially by the species *Eucalyptus regnans, Eucalyptus marginata,* and *Eucalyptus diversicolor.* They all play an important role as a source of timber.

The leaves of *Eucalyptus* species are also used economically because they contain relatively high concentrations of essential oils and oleoresins. Essential oils and oleoresins are important raw materials for the production of cosmetics and are also used for medical purposes.

The characteristic tree of the savannas in the East African mountainous regions, the baobab or monkey-bread tree *(Adansonia digitata)* (Ill. 170), is of a particularly impressive appearance. Although it does not reach the height of redwood or eucalyptus trees, when growing old it develops an extraordinarily thick trunk with heavily branched, bulky boughs. Because of their voluminosity, baobabs are sometimes called the pachyderms of the plant kingdom. One of the thick trunks is capable of storing up to 120,000 liters of water. This is a very rare phenomenon in the plant kingdom. During the drought, the trees shed their leaves. As a consequence, the giants among trees are leafless for more than six months every year. Only the gourd-like fruits hang down on long stalks. During the rains, the palmate leaves and the showy yellow flowers attaining a diameter of 12 cm appear. The fruits are edible but not particularly tasty so that the natives usually leave them to the monkeys, a fact suggested by their name.

Oak tree

Giant forms exceeding by far the usual measure are found not only among trees but also in other plant groups. For example, giant kelp *(Macrocystis pyrifera)*, a genus of brown algae growing on the temperate shores of the southern hemisphere, attains a length of up to 60 m, with a shoot diameter of 1 cm. Giant forms among grasses are a few bamboo plants whose 15-cm thick stalks grow to a height of 40 m.

Among cacti, there also are species that are impressive because of their striking size. Curt Backeberg, the famous cacti researcher and collector, called *Echinocactus grandis*, a giant cactus found in the Mexican desert of Tehuacán (Ill. 171), a saurian cactus. These reach a height of several meters and a weight of more than one ton. Nothing reliable, however, is known about their age. The juicy pulp inside the body of the cactus is edible, and has served to quench the thirst of wanderers through the desert. The peeled bodies of young specimens are used for producing candied slices of an agreeably sweet taste by boiling in cane sugar solution.

The golden ball cactus *(Echinocactus grusonii)*, native to Mexico, reaches almost the same enormous size. It attains a height of up to 1.3 m and a thickness of up to 1 m. A striking sight is the strong yellowish spines. Natives occasionally call this cactus "Seat for mother-in-law."

The giants among the bromeliads, which belong to the species *Puya raimondii* (Ill. 172), grow in the Andes of Peru in dry, rocky habitats on foggy hills, on rocky walls of the snow region, and on sunny slopes of hot valleys where rainfall is rare, in general at altitudes from 2,000 to 4,000 m. They form part of the so-called puna vegetation and are a characteristic species of the Peruvian higher Andes. There, the climate shows extreme variations; within a short time, summer heat changes over to icy temperatures. From the low-growing matted vegetation the huge trunk-forming bromeliads rise like trees. The trunk rises from a large rosette up to a height of 4 m, bearing a remarkable inflorescence that grows to a length of 5 m. This inflorescence comprises up to 8,000 flowers.

Extremely large forms, however, are found not only in plants as a whole. Individual organs of some species show extraordinary sizes. Let us take a look at some examples.

About 450 years ago, Portuguese sailors picked floating fruits out of the Indian Ocean without knowing their origin. These were double coconuts, which had a length of half a meter and a weight of up to 5 kg, thus being much bigger than normal coconuts *(Cocos nucifera)*. They were hollow and hence could not germinate. Ferdinand Magellan (1480–1521), who led the first expedition to circumnavigate the globe, reported that this remarkable nut was believed to stem from a tree growing on the bottom of the sea; therefore it was called *coco de mer* (sea coconut).

For 250 years, the native country of this giant fruit, which had reached a fantastic value through speculations, remained unknown. In 1768, an expedition landed on the island of Praslin in the Seychelle Islands. On this occasion, the surveyor Barree was the first to find a tree of the sea coconut, a lofty fan palm, in dense jungle. The French naturalist Pierre Sonnerat paid a visit to the Seychelles in 1771 and gave the newly discovered palm the scientific name of *Cocos maldivica*. Today it is called *Lodoicea maldivica* (Ill. 173, 174). In contrast to the coconut palm, it does not grow on beaches but in virgin forests.

The palm grows to a height of 30 to 40 m, and is estimated to become 600 to 800 years old. The plants grow and develop very slowly; after about 25 years they flower and bear fruit for the first time. It takes seven years for the fruits to become mature; then they have a weight of 15 to 20 kg each and fall from the trees. These nuts are the largest seeds in the plant kingdom.

The area of distribution of this palm is very small. At present, it grows only on two islands, Praslin Island, of 300 square km, and Curiense Island, of 4 square km. The total number of trees in existence is estimated at 4,000; they are protected by nature conservancy regulations.

The nuts germinate *in situ* in loose, moist humus. Fresh nuts capable of germinating cannot float because of their high density of 1.2 g/cm³; therefore, a natural spreading to other islands is impossible.

After discussing a plant with huge seeds, we shall now consider a plant that is conspicuous because of its large leaves.

This is *Victoria amazonica* (Ill. 175), a plant of the water lily family. Since it was discovered in 1801 by the botanist Thaddaeus Haenke and Father La Cueva in an estuary of the Amazon in South America, a plant has rarely received so much attention as this large floatingleaf plant. In literature, stories of repeated discoveries of this plant can be read. Haenke died during his journey, and part of his scientific work was lost.

Aimé Bonpland, Humboldt's companion, was the next to find the plant in 1819. More details on this imposing aquatic plant, however, were given by Alcide d'Orbigny, who found it in 1827 and wrote a scientific report on it in 1840. Meanwhile other scientists had also discovered this plant, in 1832 Eduard Poeppig and in 1836 Robert Hermann Schomburgk.

Victoria amazonica grows in still waters, in bays of lakes, in swamps of the Amazon region as well as in Bolivia and Guiana. Most frequently it is found in tributaries of the Amazon where they cover the surface of the water with their giant leaves over many kilometers. The thick rhizomes of the plant stand erect in the mud of the shallow waters being retained by countless roots. The circular leaves attain a diameter of up to 200 cm, and the edges, which form a tray, have a height of 4 to 6 cm. The Red Indians at the Parana rightly called this plant *irupe* (water bowl). On the upper side, the leaves are smooth, whereas the underside is provided with many spines and a system of board-like ribs connected by lateral crossbands (Ill. 176). Individual ribs are up to 6 cm high and 2 to 5 cm wide, containing air ducts. This gives the leaves a high carrying capacity. A fully grown leaf carries a load of up to 75 kg. So it could carry a grown-up man if his weight were distributed uniformly over the leaf area. As this is impossible, the occasionally published pictures showing a man sitting on a Victoria leaf must be considered as trick shots. The rate of growth of the leaves is amazing. An unfolding leaf, for example, shows an increase in size of 2.5 square cm in one hour. This corresponds to a leaf area of 0.3 m in 24 hours. In this manner, a single plant develops a leaf area of 60 square m in 20 to 25 weeks. This rate of growth is very rare in the plant kingdom.

The life of a leaf is 6 to 8 weeks. To be able to breathe and assimilate, the upper side of these giant leaves is provided with tiny holes (stomata) that allow rain water to ooze downward. About one dozen such openings have been counted on one square cm of leaf area.

Imported seeds, kept in a bottle with fresh water, germinated for the first time in the spring of 1849 in England, where the plants eventually put forth flowers. Today, *Victoria amazonica* is an attraction in almost all botanical gardens. The normally perennial plant is cultivated as an annual plant in large water basins as in temperate latitudes the lack of light in winter would cause its death. The seed is sown in January or February at high temperatures. The first flowers will then appear in June. They have a size of 20 to 40 cm and open for two nights only. The originally purely white flower opens toward the evening and closes in the morning. In the afternoon, it opens for the second time, taking a pink to dark-red tinge. Then this exotic relative of the Central European water lilies withers, and its seeds grow ripe under water. When unfolding, a strong pineapple-like scent emanates from the flower. Besides this species, the smaller *Victoria cruciana*, which occurs in regions farther to the south in Parana, Uruguay, and northern Argentina, is well known.

After discussing plants with large seeds and large leaves, let us now consider the plant with the largest flowers. It is *Rafflesia arnoldii*, which was discovered in 1818 by the botanist Arnold in dense tropical virgin forest on the island of Sumatra. Its flowers attain a diameter of one meter; some scientists report extremely large forms having a diameter of 1.4 m (Ill. 177). The huge brown-red petals, folded over toward the outside, form an open bowl in which carpels and stamens are arranged. The flowers are the only large part of this plant; all other organs are small or missing. It lives as a parasite on the roots of lianas of the genus *Cissus* and, therefore, has neither a stalk nor leaves. The flowers lie directly on the ground and only small and thin suckers penetrate into the roots of the host plant. The seeds are tiny too. A strong smell of decaying and feces emanates from the flowers. This smell and the color of the flower attract carrion flies and beetles which perform pollination when depositing their eggs in the flower.

When talking of the oldest plants, people usually think of the proverbial millenial oaks. The common oak or pedunculate oak *(Quercus robur)* may indeed grow so old. These trees are very rare, however. According to estimates by scientists, the common oak can reach a maximum age of 1,200 years. In the Ivenacker Park near Stavenhagen in the German Democratic Republic, there is on oak that is estimated to be 1,000 years old. The same age is ascribed to a 15-m-high elm at Schümshein (Rhinehesse). A relatively high age is also ascribed to the spruce *(Picea abies)*, attaining 1,000 years, and to the red beech *(Fagus sylvatica)*, attaining 900 years. These three species are among the oldest living trees in Central Europe. All species—with one exception—live considerably shorter life spans.

This exception is the yew *(Taxus baccata)*; frequently it does not look its age because it grows very slowly. Its wood is dense and heavy and the annual rings are closely spaced. Yew timber has been in great demand for a long time and, therefore, has been rooted out almost completely in many places in Central Europe. For this reason, it is protected by nature conservancy regulations in most countries.

The age of yews, especially that of larger and thicker specimens where several trunks have grown together, cannot be determined exactly. It is estimated, however, that yews grow to be more than 2,000 years old. A 3,000-year-old yew is reported to grow near Braburn in the county of Kent in England, and several trees in the yew grove of Sochi, in the Crimea, are said to be even older. This means that yews reach practically the same age as the North American redwood trees, and both of them are among the plants attaining the greatest age.

The oldest trees of all grow in the Inyo National Forest in the White Mountains of China; in the rain shadow of the Sierra Nevada in California, at altitudes of 3,000 m; and on Wheeler Peak in eastern Nevada at elevations between 3,200 and 3,700 m. It should be noted that the oldest plants, and thus the oldest living beings, are the bristlecone pines *Pinus aristata* (Ill. 179). Recent examinations have shown that trees reach the incredible age of more than 4,000 years. At least five specimens are older than 3,000 years, three are said to be more than 4,000 years, and one, 4,900 years old. In comparison with this, the 1,200 years of the common oaks are rather modest. The bristlecone pine, however, does not have an imposing appearance; it grows to a height of only 5 to 12 m.

The bristlecone pine grows on windbeaten and desert-like slopes without any undergrowth. The high resin content protects the wood from moisture and rot. The particular feature of the growth of this tree is the fact that it responds in an unusually high degree to environmental influences (Ill. 180). This is indicated by the general pattern of the annual rings. The annual growth as indicated by the annual rings is so small that it cannot be perceived without the help of a microscope. The University of Arizona possesses a wood sample 12.7 cm long on which 1,100 annual rings were counted. This means an annual increase in thickness of roughly 0.1 mm.

With the help of the annual ring chronology, proof can be given of the climate and certain events of times long past, such as years of drought and disaster—even of the earthquake that destroyed San Francisco in 1906 (Ill. 178).

Strange shapes, rare plants

Occasionally, monstrosities occur in the plant kingdom which are changes of the appearance of a plant or of individual plant organs. Such malformations of plants aroused interest in the Middle Ages, as shown by descriptions of the perfoliated head of the pot marigold *(Calendula officinalis)* written by the famous Abbess Hildegard von Bingen in the 12th century. One of the first more comprehensive books on the subject of malformations in plants, titled *Missbildungen der Gewächse*, was published by Georg Friedrich von Jaeger (1785–1866) in 1814. Goethe may have received many a suggestion from this book for his morphological studies.

Abnormal forms of shoots constitute the majority of malformations. They are commonly manifested as enlargement and flattening of sprouts or shoots in plants with stem sections which are more or less round. This peculiar feature is known as fascination.

A plant that attained ornamental value only by fascination is the cockscomb *(Celosia argentea* var. *cristata)* (Ill. 181). The velvety inflorescence of this plant of the amaranth family has the curly shape of a cockscomb, which is arranged on top of an oblong and pointed leaf. The enlargement of the inflorescence to such a degree that it took the form of a cockscomb was due to a malformation which became hereditary. The flowers develop seeds only in their lower part. This species of cockscomb was known in Europe as early as 1570. It is cultivated in white, yellow, red, violet, or varicolored forms of different heights of growth. The attractive inflorescence is fully colored as if "dyed in the wool." Probably this plant was taken from its African habitat to European gardens already in the Middle Ages.

An interesting abnormality in the red foxglove or fairy bell *(Digitalis purpurea)*, which was mentioned earlier, is the formation of peloria (Ill. 182, 184). In these plants, an actinomorphic flower is situated at the top of the inflorescence, whereas the other flowers show the dorsiventral arrangement that is typical of foxglove. The pelorian flower opens first and is considerably larger than the other flowers. In pelorias up to 30 petals have been counted growing together with adjacent petals.

The formation of peloria is considered by botanists to be an atavism of the original shape of the flower. From this floral monstrosity, they also draw the conclusion that the zygomorphous arrangement of flowers was brought about relatively late in the development of seed plants.

The reason for the development of such abnormal shapes is not yet fully understood; there are scientists who think that it is due to an increased vitality.

The causes of many other deviations from the normal are known, however. When plants are infected by diseases or infected with pests, in general, growth and further development of the plant are impaired. There are, however, a few "useful" diseases. Although we will not go into detail about diseases of weeds and other noxious plants, we shall discuss a special form of virus disease that occurs in some ornamental plants.

Such a "disease" is found in the flowering maple *(Abutilon)* (Ill. 187). Several species of this genus are infected by a mosaic virus. As a consequence, the palmate, quinquelobate to septemlobate leaves of *Abutilon striatum* cv. *thompsonii* become variegated, green-yellowish-white, and only in this way attain an ornamental effect. This disease of this perennial and decorative ornamental flowering and leaf plant has been known for a very long time. It was described for the first time in an English horticultural periodical of 1868.

It is of particular interest that this disease—i.e. this mottled coloration—can be caused to disappear when the plant is exposed to high temperature—over 36 °C—for a period of several weeks. Then the newly grown shoots are healthy; frequently the whole plant is cured after this relatively short treatment, because the virus is killed at these temperatures.

The flowering maple is not the only plant that gains in ornamental value by a viral disease. Perhaps more widely known is the example of tulips infected by a viral disease *(Tulipa gesneriana)*; this applies to the striped forms of this flower embellishing our gardens in spring (Ill. 183). The different coloration of the petals is caused by this disease.

169 Mammoth trees at the East River in California
170 A baobab *(Adansonia digitata)* in the East African
thorny savannah

171 *Echinocactus grandis* in the Mexican desert of Tehuacán
172 *Puya raimondii*, the giant among bromeliads

173 Fruits of the Seychelles nut *(Lodoicea maldivica)*, which were collected at the natural site on the Isle of Praslin

174 Cotyledon of the Seychelles nut

176 Leaf underside of *Victoria amazonica*
177 The largest flowers in the entire plant kingdom
are developed by *Rafflesia arnoldii.*

178 Study of the annual rings of bristlecone pines *(Pinus aristata)*
179 Bristlecone pines *(Pinus aristata)* in the Inyo National Park
in California
180 Old bristlecone pines frequently have a ghostly appearance.

181 Cockscomb formation in *Celosia argentea*, a species frequently cultivated as an ornamental plant

182 Normal inflorescence of the red foxglove *(Digitalis purpurea)*

183 The stripes in the tulip flower *(Tulipa gesneriana)* are caused by a viral infection.

184 Formation of peloria in the red foxglove

185 Albinotic form of the normally ruby-purple flowers of the checkered lily *(Fritillaria meleagris)*

186 Side by side with the normally flowering plants of the soldier orchid *(Orchis militaris)*, the purely white specimen catches the eye.

187 Only by a viral disease does the
flowering maple *(Abutilon)* produce an ornamental effect.

188 The *Pyrenacantha* species in the dry shrubs of Kenya develops curious stem shapes.

189 *Colpothrinax wrightii* on the Isle of Pines, Cuba
190 The royal palm *(Roystonea regia)*, native to various West Indian islands, is one of the most beautiful of all palm trees.
191 The silversword *(Argyroxiphium sandwicense)* grows only on the Haleakala Crater on the island of Maui.

192 *Idria columnaris* in Lower California
193 *Mimosa pudica* in a non-irritated state
194–195 A mimosa in an irritated state

Changes in color of flowers may be due to many other causes. Plants normally flowering blue or red all of a sudden appear with white flowers. This phenomenon, called albinism, is caused by genetic mutations—i.e. relatively permanent changes in hereditary material. The chemistry of changes in color that takes place in the plant cells is complicated because many factors may exert an influence on the pigments in the cells. From 1930 to 1940, the researchers W. J. Lawrence and J. R. Price devoted themselves to the study of these problems. The frequency of occurrence of albinism in nature is roughly estimated at a ratio of 1 : 10,000. Examples of albinism are found everywhere—e.g. in the guinea-hen flower or checkered lily *(Fritillaria meleagris)* (Ill. 185) and in the soldier orchid *(Orchis militaris)* (Ill. 186). The pure white forms of this orchid are particularly striking.

Strange and interesting forms of plants are brought about not only by changes due to external and internal causes. Many species originally have shapes worth considering.

To the south of the western Cuban province of Pinar del Rio, on the Isle of Pines, a palm grows in the sandy savanna which is amazing because of its strange appearance. In the middle of the trunks, there are barrel-like swellings. It is the palm *Colpothrinax wrightii*, which is called *palma barrigona* (belly palm) by the local people (Ill. 189). The thickening of the trunk does not appear in the juvenile stage of the plant. After a period of 10 to 15 years of growth, a secondary meristem develops in the middle of the trunk so that it thickens considerably. It is an interesting fact that the belly palm will flower and bear fruit only after the formation of the trunk thickening.

Similar swellings, but less pronounced, are found in the palm species *Acrocomia armentalis*, also indigenous to western Cuba, and in other species of this genus. As the trunk thickenings start at the same level above the ground, the time of their formation must be hereditary. The swellings are water and reserve substance stores.

The stems of the most beautiful palm, the royal palm *(Roystonea regia)* (Ill. 190), occurring wild in river valleys of the West Indian islands, are somewhat swollen in their middle parts. The smooth stems attain a height of up to 30 m and bear a crown of beautiful pinnate fronds. The panicle-shaped inflorescences appear under the fronds. The stone-fruits, which have a high oil and starch content, are used as fodder. Because of its impressive shape, the royal palm is widely used as an alley and park tree in the tropics. In Cuba it also determines the scenery of open landscapes and is found in the national emblem.

Trees of the species *Idria columnaris*, a characteristic plant of dry arid regions in Lower California (Ill. 192), show curious shapes. The arborescent trunk succulent, a plant of the candlewood family, grows to a height of 18 m. Large numbers of short and thin side branches grow out of the spongy stems, which give a bizarre appearance to the plant. From afar, the plants look like bare tapering rods. That is why they are also called "telegraph pole plants."

The stem of another plant grows more broad than high. The bulbous thickened stem of the *Pyrenacantha* species, occurring in the dry bush regions of Kenya, resembles a piece of rock on which a root grows (Ill. 188). A climbing shoot is put forth from the middle of the tuber and turns like a coil. The tubers of *Pyrenacantha malvifolia* can attain a diameter of 1.5 m when they are old. The juicy pulp is eaten by elephants and rhinoceroses.

The Hawaiian silversword *(Argyroxiphium sandwicense)*, a plant of the composite family, grows only on the volcanic crater Haleakala on the island of Maui (Ill. 191). The young plant resembles a densely leaved agave. The silvery bright leaves are densely haired and prickle-shaped. From afar, the young plants resemble silvery balls. A period of seven to twenty years elapses before the plant will form flowers. The flower stalk rises 50 cm above the silvery ball, producing more than one hundred purple sunflower-like inflorescences. After fructification, the plant dies. It is of particular interest that seven different insect pests have specialized on silversword, threatening the further existence of this rare plant.

Toward the end of the 18th and at the beginning of the 19th century, ruthless souvenir hunters almost eradicated this plant because of its rareness and beautiful color. Today the region of Haleakala is a nature preserve. A vast crater produced a landscape of volcanic cones and other formations. Vegetation is scarce, but the charm of the scenery is unique.

Plants attract attention not only because of their unique shapes and particular colors but also because of certain other properties found in some species.

There is a saying, "as sensitive as a mimosa." Indeed, the sensitive plant *(Mimosa pudica)* responds, in ways that are unusual in the plant kingdom, to vibrations, touching, and shock (Ill. 193, 194), and the mimosa has become a popular illustration of this phenomenon, which is known as seismonastic motion. When a leaf is touched, the small pinnules turn upward, and the leaf stalks or petioles lower in the direction of the stem of the plant. The irritation is transmitted in the plant over a distance of 50 cm at a rate of up to 10 cm per second. After 10 to 20 minutes the irritation fades away, and the leaves assume their original position (Ill. 195). The motion is caused by changes in the turgor pressure in certain cellular tissues. When cell sap enters the cellular interstices, the conditions of tension of the individual parts and the whole tissue are changed, producing the flapping motions.

The mimosa is indigenous to Brazil, but it now grows wild in many tropical countries. In hothouses it is cultivated as an annual plant. The upright stalks grow to a height of 80 cm and are somewhat prickly. They bear the small dark-green pinnules. The pink-white to light-red flowers are arranged in spherical inflorescences.

At the end of this chapter and of the book, let me introduce to you a citrus plant unique throughout the world that grows on the soil of the Test Station for Subtropical and Southern Cultures in the Black Sea spa of Sochi in the Crimea. Neither its age nor its size is the interesting aspect of this "Tree of Friendship." It is only 40 years old and not higher than 4 m (Ill. 196). Since 1940, when the Soviet Arctic explorer Otto Schmidt, staying at the test station as a guest of honor, grafted this tree for the first time, the citrus tree has been subjected to many grafts that were carried out by representatives of more than 130 countries, by people of different nationalities and professions. Fresh scions were grafted on the branches of this tree by Yuri Gagarin, Paul Robeson, Ho Chi Minh,

Marshal Voroshilov, and many other personages. Today, more than 45 different citrus fruits grow in the dense crown of the tree. Beside Japanese mandarins, there are Italian lemons and American grapefruits; beside small kumquat fruits hang very large oya fruits. Near the fruits, small white labels with names in many languages are arranged. Before a guest is allowed to apply the grafting tool to the Tree of Friendship, he is taught the basic rules of grafting.

Here we end our incursion into the wealth of variety that is the vegetable kingdom. There are more than 370,000 kinds of plant life on earth; in this book we could deal only with a small selection. Selected examples have demonstrated remarkable cases, and other, less immediately noticeable peculiarities, and have been used to reveal to the reader the infinite variety and beauty of plants, and thereby encourage him to seek further in this fascinating realm.

Glossary

Index of plant names

Appendix

Bibliography

Sources of illustrations

Glossary

Alkaloids

An important group of alkaline natural substances that are formed in the so-called secondary metabolism of plants. They are nitrogen ring compounds of different chemical structures, which combine with acids without elimination of water. Usually they are formed in roots, stored in roots, barks, leaves, and fruits. Many of them are poisons or are used as drugs because they act on the nervous system (analgesic, antispasmodic, anesthetic).

They are found in plants of the poppy, nightshade, and other families, with several alkaloids occurring in one plant. Well-known alkaloids are ergotamine, ergotoxine, ergometrine, hyoscyamine, atropine (deadly nightshade), aconitine (aconite), chinine, morphine, codeine, papaverine, coniine (poison hemlock), strychnine, cocaine (coca), nicotine (tobacco), caffeine (coffee, tea), and colchicine (meadow saffron).

Alternation of generations

Regular alternation of a sexual and an asexual generation, which is particularly pronounced in cryptogamous plants. In these plants, the sporophyte, which produces asexual spores, alternates with the gametophyte, which bears germ cells, the gametes. This change is usually connected with a change in the chromosome number. Alternation of generations is typical of ferns, where spores are formed on the sporophyte, the fern plant body proper; and from the spores, the gametophyte or prothallium, the sex organs originate. After fertilization, a new sporophyte develops from the zygote. Due to a reduction of the gametophyte, the alternation of generations has become indistinct in flowering plants.

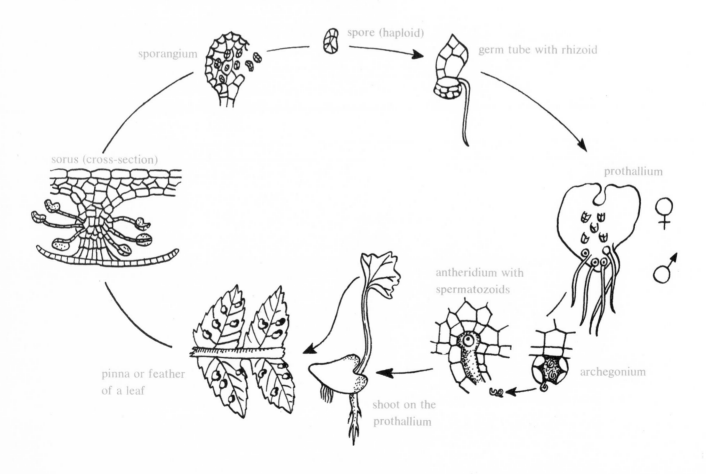

sporangium

spore (haploid)

germ tube with rhizoid

sorus (cross-section)

prothallium

♀

antheridium with spermatozoids

♂

pinna or feather of a leaf

shoot on the prothallium

archegonium

annual

A plant that completes its life cycle in one growing season. The period from seed germination to the formation of new seeds covers not more than one year. The unfavorable season of the year is passed in the form of seeds only.

Annual ring

The layer of wood produced in a tree by the growth of a single year, but varying in thickness depending on the season of the year. In spring, when the rapid growth of the leaves causes the conduction of large amounts of water, the trees form the large cells of the soft early wood. Later the cells become smaller and smaller so that the harder late wood is produced; after dormancy in winter, new early wood will grow, marking the previously produced ring by a distinct line. The age of trees can be determined on the basis of the annual rings.

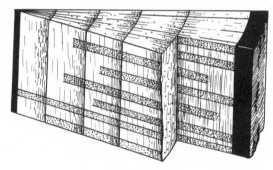

Aphrodisiac

A substance exciting sexual desire.

Area

Area of distribution of a systematic unit of plants (species, genus, family, etc.) whose boundaries cannot be crossed by the natural means of distribution (seeds, spores, runners, etc.). Boundaries of the areas are, for example, mountains, seas, and living conditions that do not agree with the plant in question. The boundaries of an area are not fixed forever, but they change in the course of the development of plants. A distinction is made between two main types of areas. In the continuous or closed area, the individual habitats of plants are so close side by side that a natural distribution from one site to another is possible. In the disjunctive area, the sub-areas are very remote from one another so that no exchange can take place. Mostly, disjunctions—that is to say, areas of discontinuity between areas in which populations of a specified organism are present—historically developed from originally continuous areas because plants of certain species died out in the areas between them. A well-known example of plants with disjunctive areas is the genus of tulip-trees *(Liriodendron)* of which one species is in East Asia, another one in the Atlantic North America. The poison ivy *(Rhus toxicodendron)* has the same distribution.

arid

Adjective used to designate an area or region as excessively dry where the possible evaporation of water is greater than the sum of rainfall. The lack of water thus produced leads to the formation of deserts or semideserts.

Assimilation

The incorporation of foreign material into the plant organism. Assimilation in a narrower sense is defined as the formation of

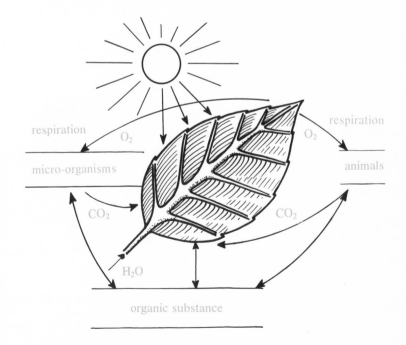

carbohydrates from carbon dioxide of the air and a source of hydrogen, as the water of the soil, exposed to light; in the first phase of this process usually glucose is formed. This means that assimilation is the basis of any life on earth.

The opposite: dissimilation, destructive metabolism involving the release of energy. Dissimilation processes are respiration and enzymosis.

autotrophic

Plants that are capable of nourishing themselves exclusively on inorganic substances and building up organic substances of them are called autotrophic. The opposite: heterotrophic.

Bastard

A hybrid, an individual produced by union of gametes from parents of different genotype. Hybrid is the more widely used term; a distinction is made between hybrids obtained by the cross-breeding of strains (within one and the same species), of different species, and different genera. The formation of hybrids plays an important part in plant breeding, especially in the case of some ornamental plants.

Bloom

A biological-functional pollination unit for attracting the animals that perform pollination and for ensuring pollination. A bloom can consist of a single flower, individual floral parts, more than one flower and other organs in addition.

Carpels

The female parts of a flower forming the ovary and bearing the ovules; see *Flower*.

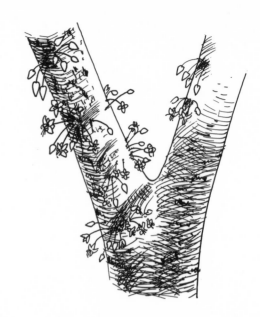

Cauliflory

The capability of producing flowers and fruits directly from the main stem or older branches.

Cephalium

A tuft of some cacti consisting of dense wool and fine bristles, borne on top, which serves for the formation of flowers and fruit.

Chitin

A highly resistant, high-molecular-weight organic nitrogen compound (glucosamine polysaccharide), which is found primarily in the skeleton of arthropods and, in very small quantities, in other animals and in some kinds of mushrooms.

Chlorophyll

The green coloring material of plants, chemically complicated, which is essential to photosynthesis. Usually it occurs in discrete bodies, known as grana, that make up most of the substance of plant chloroplasts and that are the actual seats of the chlorophyll.

Copula

Copulation; transfer of male germ cells into the female reproductive duct.

Corona

Appendage borne on the inner side of the corolla adjacent to the stamens in certain flowers, often resembling an additional whorl of the perianth, adding to the showiness. Coronas are found in many ornamental plants.

Cosmopolite

Species of plant that is distributed all over the world and whose area covers large parts of the earth.

Cuticle

A thin continuous film on the outer wall of the epidermis in plants, consisting of cutin, a waxlike substance that is practically impermeable to water and gas. The cuticle, above all, protects the plant from water losses and, therefore, is particularly thick in many xerophytes.

Cycle of substance

All substances taken up by living organisms pass through certain, sometimes very complicated cycles which include the various organisms, water, air, and soil. These cycles in essence can be divided into two sections. The organic substances are formed by the autotrophic green plants; after various processes of modification, they are decomposed into inorganic substances by the heterotrophic organisms by way of rotting and decaying. Known cycles of substances are the carbon cycle and the nitrogen cycle, among others.

dioecious

Said of a seed plant—i.e. having unisexual male or female flowers borne on different individuals. Thus, cross-pollination is recurring for propagation.

Dripping tip

The long and strengthened tip of a leaf, found in all plants of tropical rain regions, which facilitates the running off of water and thus prevents injuries during heavy rains.

Ecology

A branch of science concerned with the interrelationship of organisms or groups of organisms and their environments, including the totality of animate and inanimate factors.

endemic

A plant or group of plants that is native to a relatively small area. There are endemic species, genera, families, etc.

Enzyme

A biocatalyst, an active substance that is produced by living cells which enables or accelerates processes of chemical transformation and itself does not undergo marked destruction. Enzymes are complex proteinaceous substances of which many are present in one cell because they regulate all metabolic processes. They have different structures and excel in a great substrate specificity; i.e., they are active in one chemical transformation only.

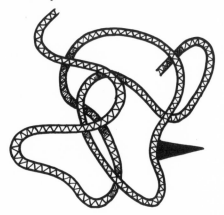

Essential oils

Complicated mixtures of organic compounds formed in protoplasm which are excreted from glandular hair of flowers, sprouts, and leaves or deposited in oil cells or oil reservoirs. In contrast to the fatty oils, they evaporate completely and do not leave fat stains on paper. Probably they are metabolic waste products which, due to their typical smell, also serve for attracting insects or for intimidation. About 150 essential oils are utilized, e.g. as odorous substances (e.g. jasmine oil, spike-oil, attar of roses), as aromatic substances (e.g. oil of bitter almonds, nutmeg oil, caraway seed oil, cinnamon oil), or for the production of drugs (e.g. fennel oil, camomile oil, camphor oil). The essential oils are won by water-steam distillation, pressing out or extraction with the help of various solvents (benzene, petroleum ether, and others).

fertile

capable of reproducing.

Filament

The anther-bearing stalk of a stamen; see *Flower.*

Flora

The total number of plant species existing in a certain area of the earth or during a geologic age.

Flower

The transformed shoot serving for sexual reproduction, usually of limited growth. A complete flower of the angiosperms consists of the torus, the sepals, petals, stamina, and carpels, and sometimes of special devices such as nectaries. Most floral parts are transformed leaves. The stamina consist of the filament and the anther, in which pollen are formed. The carpels form the ovary, with the style and the stigma. One or more carpels can be included in the formation of the ovary. The ovary contains one or several ovules that develop into seed. Due to concrescence or reduction, individual floral parts may be missing, which may lead, for example, to the formation of unisexual flowers. In general, reduced flowers are considered to be derived phylogenetically.

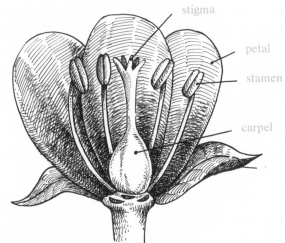

Form of life

Plant forms developed by a similar way of adaptation to the surroundings—e.g. in growth or in their overwintering organs—but that can be included in different groups of relationship. Sprouting plants can be grouped in five forms of life—i.e. the phanerophytes, including the trees, shrubs, climbers, and epiphytes; the chamaephytes; the hemicryptophytes; the geophytes; and the therophytes.

Fossil

Any petrified remains, impressions, or traces of organisms that lived in past geological ages.

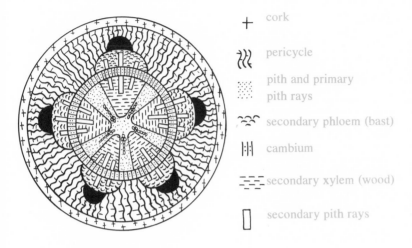

- cork
- pericycle
- pith and primary pith rays
- secondary phloem (bast)
- cambium
- secondary xylem (wood)
- secondary pith rays

Growth, secondary

Growth in higher plants that results from the activity of a cambium—i.e., a formative layer between xylem and phloem—producing increase in diameter and supplying supporting and conducting tissue. Secondary growth takes different forms in conifers, in monocotyledons, and in dicotyledons.

Haustoria

Food-absorbing outgrowths of parasitic or semiparasitic plants which penetrate the conductive tissue of their host plants from which they take nutritive substances and water. Usually these are modified roots or parts of roots.

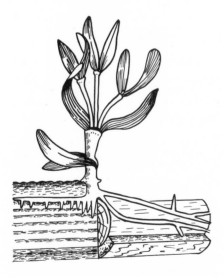

heterotrophic

Organisms that obtain nourishment in the form of organic substances from outside sources because they are not capable of producing them are called heterotrophic. Heterotrophic organisms include animals and most of the non-green plants (mushrooms, full parasites, and others). Opposite: autotrophic.

Humus

The organic portion of soil formed by the partial decomposition of vegetable and animal matter and microorganisms imparting the characteristic color to the soil. Complex chemical processes of transformation take place in the humus; it plays an important part in soil fertility. Depending on the initial material and the degree of decomposition, a distinction is made between various forms of humus—e.g. raw humus, rot, and mull.

Inflorescence

A floral axis with its appendages, part of the shoot, frequently much branched and bearing many flowers, not bearing true leaves but only bracts or spathes. The inflorescence is clearly distinguished from the region of leaves. According to the mode of arrangement of flowers, inflorescences are classified, for example, as raceme, ear, spadix, catkin, panicle, umbel, and capitulum.

monoecious

Having unisexual male and female flowers in one and the same plant. This precludes self-pollination, but geitonogamy is possible.

multi-annual

A plant which requires several years for its development, flowers once, and dies after flowering. Plants flowering several times are called perennials.

Mycelium

The mass of interwoven hyphae that forms the vegetative portion of a fungus. The mycelium grows especially below ground surface. The fruit bodies rise from the mycelium and above ground level.

Nectar

An aqueous solution of various sugars and other substances, especially odorcus substances, that is secreted by the nectaries of a flower or other plant organs and that serves for attracting the animals for pollination.

Nectaries

Gland surfaces or glandular hair secreting nectar in a flower (see *Flower*). In simple flowers, nectaries are relatively free and easily accessible for the pollinators. In derived flowers pollinators frequently have to pass through long appendages or the like, so that pollination is ensured.

Nomenclature, binary

A scientific method for the designation of plants developed by Carl von Linné, used by him for the first time in his work *Species plantarum* in 1753. According to this nomenclature, the name of a plant species consists of two parts, of which the first is the genus, with a capital letter, and the second is the species itself—e.g., *Tilia platyphyllos,* broad-leaved lime tree, *Tilia cordata,* small-leaved lime tree. Today international rules are generally accepted for the designation of plants.

Ovary

See *Flower.*

Ovule

Also known as seedbud; the female organ of reproduction in seed plants; see *Flower.*

Paleobotany

A branch of botany that deals with fossil plants.

Parasite

An organism living in or on another living organism, obtaining from it nutriment without offering any advantage to the host. Parasitic plants are characterized by reduced leaves and the absence of chlorophyll. In higher plants, which usually are connected with their host by haustoria, a distinction is made between sprout parasites such as the boxthorn or matrimony vine *(Cuscuta)* and root parasites such as the broom rape *(Orobanche).*

Perianth

The external envelope of a flower consisting of leaves (petals), which are not differentiated into calyx and corolla. The perianth may be corona-like and colored as in the tulip *(Tulipa),* or calyx-like and green, as in the nettle *(Urtica).*

Petals

A leaf-shaped member of the corolla; see *Flower.*

Photosynthesis

A basic process of life; formation of carbohydrates from carbon dioxide of the air and water from the soil with the help of chlorophyll, taking advantage of the light energy; only green plants are capable of this process.

Photosynthesis takes place according to the following general formula

$$6CO_2 + 6H_2O + \text{light energy} \rightarrow C_6H_{12}O_6 + 6O_2$$

It can be divided into three phases: namely, the light adsorption by chlorophyll, a photochemical release of oxygen through the decomposition of water, followed by combining and reducing the carbon dioxide into grape sugar. In this process, one square meter of leaf surface produces about one grain of sugar per hour; this means that all green plants combine about 100,000 million tons of carbon every year. This is equal to one hundred times the world's production of coal.

Pollen

The fine dust appearing in a flower; see *Flower.*

Pollination

The transfer of pollen from a stamen to the stigma of an ovary or, in gymnosperms, directly to the ovule; the precondition for fertilization. In self-pollination, pollen is transferred from the anther of a flower to the stigma of the same flower; in the case of geitonogamy, pollen is transferred from one flower to another growing on the same plant; and in the case of allogamy or cross-pollination, pollen is transferred from one flower to the stigma of another flower on a different plant of the same species. As in the majority of plants cross-pollination promises to be most successful, many forms of adaptation of the plant were developed to prevent self-pollination—e.g. by self-sterility, different times of maturity of pollen and stigma of one and the same flower, unisexual flowers or plants.

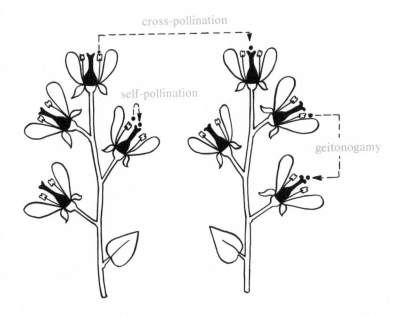

Pollinium

Also known as pollinarium; the coherent mass of pollen grains which are transferred as a unit in pollination. Pollinia are formed by orchids and in some other plant families.

Prickles

Pointed, strong modified processes arising from the upper cellular layers or from the epidermis of plants. In contrast to prickles, spines are modified plant organs.

recent

Of the present time; in contrast to fossil.

Rhizome

Rootstock, the underground, more or less thickened part of the sprout axis which is distinguished from a true root by buds, nodes, and usually scale-like leaves. Rhizomes function as organs for the storage and preservation of food material and for vegetative reproduction.

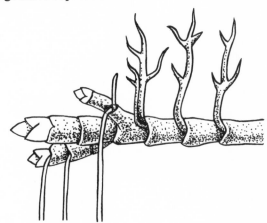

Saprophyte

A plant living on dead or decaying organic matter. Saprophytes include many bacteria and fungi which produce an important precondition for many cycles of substances (e.g. carbon cycle, nitrogen cycle).

Semiparasites

Either autotrophic or heterotrophic plants, which take certain substances, especially water and mineral substances, from their hosts to which they are connected mostly by roots. As they usually are capable of assimilating, semiparasites have green leaves. Examples are mistletoe, eyebright, cowwheat, and rattlebox.

Sepals
Leaves of the calyx; see *Flower.*

Spine
A stiff sharp-pointed plant process, branched or nonbranched, as a modified leaf or other plant organ. Depending on the organ from which it originates, a distinction is made between shoot spine, leaf spine, stipule spine, root spine and other spines; they are frequently found in plants in arid regions, and they also serve as a protection against animals trying to feed on them and as a means for retention in climbers.

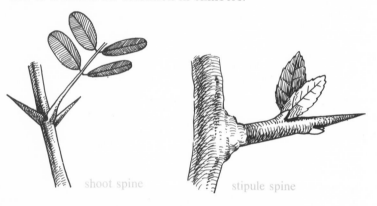

shoot spine   stipule spine

Spores
Elementary cells for asexual reproduction that develop without being fertilized. Usually they are formed in cases, the sporangia, which frequently are found on special spore-bearing leaves, the sporophylls which are distinguished from the assimilating leaves. In many plants, such as the ferns with a sexually differentiated prothallium, the spores can be differentiated into male androspores and female gymnospores from which the different gametophytes (see *Alternation of generations*) arise.

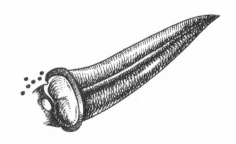

Sporophyll
See *Spores.*

Stamens
See *Flower.*

sterile
Incapable of producing offspring, incapable of germinating.

Stigma
See *Flower.*

Stomata
Minute openings in the epidermis of leaves, stems and other plant organs through which gaseous interchange between plant and air takes place—that is to say, through which water vapor and oxygen are emitted and carbon dioxide is taken up. The stomata consist of two chlorophyll-containing closing cells, between which the central opening is arranged and the adjacent, usually nongreen, secondary cells are found. One square centimeter of leaf surface can accommodate up to 700 stomata, which open and close by turgor motions.

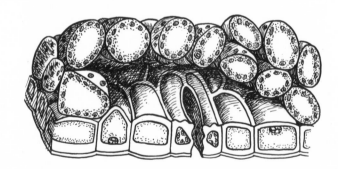

Style
See *Flower.*

Succulents
Plants developed by adaptation to particularly dry habitats that are capable of storing large amounts of water in modified organs. Succulents usually show other results of adaptation such as reduction of leaves, approximate spherical shape, and matted hairy growth on the outer surface as a protection against evaporation of water. Mainly, a distinction is made between leaf succulents, such as plants of the orpine family—aloe and agave—and stem succulents such as cacti and various plants of the spurge family.

Symbiosis
The living together of two dissimilar organisms or organisms of different species for mutual benefit, which can be so intimate that the two organisms appear to be a unit. Examples of this are lichens consisting of fungi and algae. The mycorrhiza is the symbiosis of the mycelium of a fungus with the roots of higher plants, especially trees; as a consequence, certain species of fungi grow close by certain trees (rough boletus: birch, yellow boletus: pine). Symbiosis may also be defined as mutual parasitism.

Turgor
The distention of the wall of a cell caused by the fluid content which strengthens the plant body provided water supply and consumption are well balanced. Lack of water reduces the turgor, so that the plant will droop and wilt. The opening and closing of stomata is due to different turgors in the closing cells.

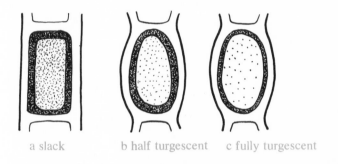

a slack   b half turgescent   c fully turgescent

Vegetation
The total plant cover of the earth or of a certain region produced in interaction between stock of species and their surroundings; totality of plant societies.

Virus
Minute particles of different shapes and structures consisting of protein bodies (nucleoproteins). They are capable of growth and multiplication only in living cells. However, they are crystallizable; thus, they are at the boundary between living and inanimate matter. The viruses include important causative agents of diseases in man, animal, and plant—e.g. smallpox, influenza, rabies, foot-and-mouth disease, and tobacco mosaic.

Witches' broom

A disease in plants caused by parasitic fungi, especially those of the genus *Taphrina*; it leads to a tufted growth of small branches on a tree or shrub.

Witches' ring

Annular arrangement of the fruiting bodies of fungi, usually about a tree. The formation of this ring occurs because the mycelium grows toward the outside while the inner parts die away.

Zygote

A fertilized egg cell.

Index of plant names

Unitalicized numerals refer to the text,
numerals in italics refer to the numbers of the illustrations.

Abies alba 159, 171
Abutilon striatum cv. *thompsonii* 176, *187*
Acacia spirocarpa 14
Acanthosicyos horridus 12, *8*
Achillea millefolium 139
Aconitum napellus 135, 138, *142*
Acrocomia armentalis 193
Adam's needle 65, *88*
Adansonia digitata 14, 172, *170*
Adansonia grandidieri 36, *45*
Adenium obesum 14
African hemp 35, *37*
Aleppo pine 33
Allium 105
Allium paradoxum 105
Allium sativum 135
Allium scodoprasum 105
Alluaudia procera 36, *43*
Aloe dichotoma 12
Aloe pillandsii 12
Alpine rose 40, *53, 54*
Amaryllis 35
Amorphophallus eichleri 61, *67*
Amorphophallus titanum 61
Angraecum sesquipedale 63, *74*
Anise 139
Ant plant 96, *109, 110*
Araucaria 158
Arbor vitae 159
Arbutus andrachne 33
Arbutus unedo 33, *26*
Argyroxiphium sandwicense 193, *191*
Aristolochia brasiliensis 61, *71*
Aristolochia elegans 61, *70*
Aristolochia grandiflora 61
Aristolochia lindneri 61
Arnica 139
Arnica montana 139
Artemisia maritima 100

Arum maculatum 60, *66, 68*
Atropa bella-donna 135
Avicennia nitida 99, *118*

Bald cypress 98, 158, 159, *117*
Banksia coccinea 38, *48*
Baobab 14, 172, *45, 170*
Bee orchid 65
Belly palm 193, *189*
Bertholletia excelsa 11
Betula pubescens 39, 98
Bird-of-paradise flower 67, *93*
Bird's-nest orchid 107, *135*
Birthwort 60
Bittersweet 95
Black saxaul 16
Bog bilberry 98
Brachystelma barberiae 59, *64*
Bristlecone pine 175, *178, 179, 180*
Broomrape 108
Brugiera 99
Bryophyllum 105, *127*
Butterwort 102, 103, 104, *121*

Calamus 11, 95, *102*
Calendula officinalis 176
Calluna vulgaris 98
Caltha palustris 68
Canary Island date palm 34
Canary Island laurel tree 34
Canna 11
Cannabis indica 143
Cannabis sativa 143, *151*
Caoutchouc liana 11
Caprifig 65
Caraway 135
Carnegia gigantea 14, 15, *20*
Carob tree 33
Carrion flower 60
Carum carvi 135

Casuarina equisetifolia 37, 47
Catasetum callosum 63, *76*
Catchfly 68
Cattleya 11
Celosia argentea var. *cristata* 176, *181*
Centaurium minus 135
Centaury 135
Cephalocereus senilis 15, *17*
Ceratonia siliqua 33
Ceropegia cancellata 62
Ceropegia distincta var. *haygarthii* 62, *73*
Ceropegia galeata 62, *72*
Ceropegia sandersonii 62
Chamaerops humilis 33, *31*
Chamomile 139, *145*
Checkered lily 193, *185*
Chicory 68
Chinese rose 67, *92*
Cichorium intybus 68
Cicuta virosa 139
Cirrhopetalum medusae 80
Cissus discolor 95, *100*
Cistrose 33
Cistus monspeliensis 33
Cistus salviaefolius 33
Cladonia rangifera 39
Claviceps purpurea 139, *144*
Clivia 35
Cluster pine 33
Cockscomb 176, *181*
Cocoa tree 11, *4*
Coconut 173
Cocos maldivica 173
Cocos nucifera 173
Colchicum autumnale 139, *149*
Colpothrinax wrightii 193, *189*
Coltsfoot 68, 139
Common bladderwort 104
Common oak 175
Common pine 39, 171
Conium maculatum 139, *146*
Conophytum 14
Corallorhiza trifida 107, *134*

Coral root 107, *134*
Coral tree 67, *90*
Cork oak 33
Corn poppy 68
Costus speciosus 59, *60*
Cotton grass 98, *116*
Coussapoa dealbata 11, *2*
Cowweed 108, *137*
Cranberry 68, 98
Crowberry 98
Cuckoopint 60, *66, 68*
Cucurbita pepo 68
Cuscuta epilinum 108
Cuscuta europaea 108
Cuscuta trifolii 108
Cylindropuntia bigelowii 22
Cypripedium calceolus 62, *69*

Dactylis glomerata 105
Daffodil 33
Date palm 12, *5*
Datura stramonium 135
Datura suaveolens 68, *95*
Dawn redwood 159, *164*
Deadly nightshade 135
Decabelone grandiflora 59, *59*
Dendrobium 11
Dentaria bulbifera 105
Desert rose 14
Devil's milk 135
Digitalis purpurea 139, 176, *182, 184*
Dionaea muscipula 102, 104, *125*
Diphasium alpinum 40, *57*
Disa uniflora 35, *41*
Dodder *133*
Dorstenia 59, *65*
Douglas fir 159
Dracaena draco 34, *33*
Dragon tree 34, *33*
Drosera anglica 102
Drosera capensis 120
Drosera intermedia 102
Drosera rotundifolia 102
Dryas octopetala 39, *52*

Dwarf juniper 40
Dwarf palm 33, *31*
Dwarf pine 40

"Ear-drops of the princess" 67, *91*
Echinocactus grandis 173, *16, 171*
Echinocactus grusonii 173
Echium wildpretii 34, *32*
Edithcolea grandis 59, *58*
Elaeis guineensis 11
Elephant grass 14
Empetrum nigrum 98
Encephalartos caffer 161
Encephalartos transvenosus 159
Enchanter's nightshade 35
Epilobium angustifolium 68
Epipogium aphyllum 64, *79*
Epuisetum giganteum 155
Equisetum telmateja 155, *154*
Ergot 139, 140, *144*
Erica arborea 33
Eriophorum vaginatum 98, *116*
Erodium cicutarium 68
Eryngium viviparum 105
Erythrina crista-galli 67, *90*
Erytropappus 35
Eucalyptus diversicolor 38, 171, *50*
Eucalyptus marginata 171
Eucalyptus regnans 171
Eucalyptus tree 37, 171
Euphorbia canariensis 34, *34*
Euphrasia officinalis 139, *138*
Evening primrose 68
Evergreen sequoia 171
Eyebright 108, 139, *138*

Fagus sylvatica 159, 175
Fairy bell 176
Fenestraria 14
Ficus bengalensis 11, *112*
Ficus carica 65, *87*
Fig tree 65, *87*
Fir 159

Fire weed 68
Flowering maple 176, *187*
"Flowering stones" 14, *12, 13*
Fly orchid 84
Foeniculum vulgare 139
Four-leaved clover 135
Foxglove 139
Freesia 35
Fritillaria meleagris 193, *185*

Galium odoratum 135
Garlic 105, 135
Giant horsetail 155, *154*
Giant kelp 173
Gibbaeum 14
Ginkgo biloba 157, *158*
Ginseng 142
Glasswort 100, 101, *119*
Glechoma hederacea 135
Golden ball cactus 173
Granadilla 60
Grape hyacinth 33
Green-winged orchis 137, *148*

Haloxylon aphyllum 16
Haloxylon persicum 16, *24*
Hammarbya paludosa 64
Heather 98
Hedera helix 95
Hemlock 139, *146*
Henbane 135
Heron's bill 68
Herschelia graminifolia 35, *42*
Herschelia purpurascens 35, *39*
Hevea brasiliensis 11
Hibiscus regius 66, *89*
Hibiscus rosa-sinensis 67, *92*
Hibiscus schizopetalus 67, *91*
Holly oak 33
Hoodia bainii 14, *7*
Horsetail 155
Hop 95
Huernia hystrix 14, *10*
Huernia zebrina 14, *9*
Humulus lupulus 95

Huperzia selago 155, *153*
Hyoscyamus niger 135
Hypericum perforatum 135, *141*

Idria columnaris 193, *192*
Indian hemp 143, *151*
Indian pipe 108
Insect orchis 33
Ipomoea purpurea 68
Ivy 95, 135

Juniperus sibirica 40

Kalanchoe tubiflora 105
Kermes oak 33
Kigelia pinnata 14, *19*
Knotweed 105

Lady's slipper 62, 63, 64, *69*
Lanceolate plantain 139
Landolphia 11
Languncularia 99
Larix sibirica 39
Lathraea squamaria 108, *139*
Laurus azorica 34
Lavandula stoechas 33
Lavender 33
Ledum palustre 98, *115*
Lepidodendron 155
Leucadendron argenteum 35, *35*
Levisticum officinale 135
Limonium vulgare 100
Lithops aucampiae 14, *13*
Lithops meyeri 14, *12*
Lobelia deckenii 40, *56*
Lodoicea maldivica 173, *173*
Lophophora williamsii 143, *152*
Loranthus europaeus 106
Lousewort 108, *136*
Lovage 135

Macrocystis pyrifera 173
Magnolia grandiflora 165, *166*
Magnolia stellata 167
Maidenhair tree 157, 158, 159, *158*

Malaxis monophyllos 64
Malayan flower 68
Mammoth tree 158, 159, 171, *169*
Mandragora vernalis 136, *143*
Mandrake 136, 137, *143*
Mangrove 98, *118, 126*
Marcgravia evenia 67, *94*
Marsh marigold 68
Marsh tea 98, *115*
Marshwort 98
Masdevallia erythrochaete 82
Masdevallia veitchiana 83
Mastic tree 33
Matricaria chamomilla 139, *145*
Meadow saffron 139, 140, *149*
Melampyrum nemorosum 108, *137*
Melocactus acunai 16, *21*
Mesembryanthemum 35
Metasequoia glyptostroboides 159, *164*
Metroxylon 11
Microcoelia physophora 81
Microcycas calocoma 160
Milfoil 139
Mimosa 194, *193, 194, 195*
Mimosa pudica 194, *193, 194, 195*
Mistletoe 106, 107, 108, *131, 132*
Monkey-bread tree 172
Monkshood 135, *142*
Monotropa hypopitys 108
Monstera 95
Mother-of-thyme 135
Mountain pine 98
Musa 67
Muscari neglectum 33
Myrmecodia echinata 96, *109, 110*
Myrtales 99
Myrtle 33
Myrtus communis 33

Nara pumpkin *8*
Narcissus poeticus 33
Neottia nidus-avis 107, *135*
Nepenthes madagascariensis 103, *122*

Nepenthes x mixta 104, *123*
Nerium oleander 33
Neuropteris 156
Nut tree 11
Nux vomica 141, *150*

Odontoglossum rossii 68, 98
Oenothera biennis 68
Oil palm 11, 33
Old-man cactus 15, *17*
Olea europaea 33
Oleander 33
Olea verrucosa 35
Oncidium cruentum 68
Ophrys apifera 65, *86*
Ophrys holosericea 65
Ophrys insectifera 64, *84*
Ophrys scolopax 33, *28*
Ophrys speculum 33, *27*
Ophrys sphegodes 65, *85*
Opium poppy 143
Orchard grass 105
Orchis militaris 193, *186*
Orchis morio 137, *148*
Orchis purpurea 64, *77, 78*
Orobranche 108
Osmunda regalis 156, *155*
Osteospermum 35
Oxalis deppei 135
Oxalis europaea 68

Pachypodium lameri 36, *46*
Pachypodium namaquanum 14, *11*
Paeonia officinalis 135, *147*
Panax shin-seng 142
Papaver rhoeas 68
Paphiopedilum villosum 68
Pará rubber tree 11
Passiflora quadrangularis 60, *63*
Passiflora racemosa 60, *62*
Passion flower 59
Pedicularis palustris 108, *136*
Pedunculate oak 171, 175
Pelecyphora aselliformis 15, *18*
Pennisetum benthami 14

Peony 135, *147*
Peyotl 143, 144
Phalaenopsis amabilis 68, *97*
Phaseolus 95
Philodendron 95
Phoenix canariensis 34
Phoenix dactylifera 12, *5*
Picea abies 159, 171, 175
Picea schrenkiana 39, *55*
Pilewort 105
Pimpinella anisum 139
Pine 33, *25*
Pinguicula alpina 103
Pinguicula vulgaris 103, *121*
Pinus aristata 175, *178, 179*
Pinus cembra 39, *54*
Pinus halepensis 33
Pinus mugo 40, 98
Pinus pinaster 33, *25*
Pinus pinea 33
Pinus strobus 159
Pinus sylvestris 98, 171
Pistacia lentiscus 33
Pitcher plant 103, *122*
Plantago lanceolata 139
Platycerium 96, *107, 108*
Pleiospilos bolusii 14, *14*
Poa alpina 105
Polygonum viviparum 105
Pot marigold 176
Protea barbigera 35, *38*
Protea cryophila 36
Protea cynaroides 35, *40*
Protea grandiflora 35
Pseudotsuga menziesii 159
Pumpkin 68
Purple orchis *77, 78*
Puya raimondii 173, *172*
Pyrenacantha malvifolia 193, *188*

"Queen of the night" 68, *96*
Quercus coccifera 33
Quercus ilex 33
Quercus robur 171, 175
Quercus suber 33

Rafflesia arnoldii 174, *177*
Ranunculus ficaria 105
Rattle 108, *140*
Ravenala madagascariensis 36, *44*
Red beech 159, 175
Red foxglove 176, *182, 184*
Red saxifrage 39, 40, *51*
Reindeer moss 39
Rhinanthus serotinus 108, *140*
Rhizophora conjugata 126
Rhizophora mangle 118
Rhododendron ferrugineum 40, *53,*
54
Ritterocereus hystrix 16, *21*
Rockfoil 105
Rockrose 33
Rosemary 33
Rosmarinus officinalis 33
Rotan palm 11, 95, *102*
Royal fern 156, *155*
Royal palm 193, *190*
Roystonia regia 193, *190*
Russian thistle 100

Saguaro cactus 14, *20*
Saint-John's wort 135, 138, *141*
Salicornia europaea 100, *119*
Salix glabra 40
Salix retuculata 40
Salsola kali 100
Sandleek 105
Sansevieria 34
Sarracenia purpurea 103, *124*
Sausage tree 14, *19*
Saxifraga nivalis 105
Saxifraga oppositifolia 39, *51*
Sciadopitys verticillata 158, *162*
Sea blite 100
Sea holly 105
Sea pink 100, *101*
Selenicereus grandiflora 68, *96*
Selenicereus macdonaldiae 68
Sequoiadendron giganteum 171, *169*
Sequoia sempervirens 171
Serapias cordigera 30

Serapias neglecta 29
Seychelles nut 173, *173, 174*
Siberian larch 39
Sidesaddle flower 103
Sigillaria 155
Silene noctiflora 68
Silkweed 108
Silver fir 171
Silversword *191*
Silver tree 35, *35*
Snowball protea *36*
Solanum dulcamara 95
Soldier orchid 193, *186*
Sonneratia 99
Sparmannia africana 35, *37*
Spathodea campanulata 11, *3*
Sphagnum 97, *114*
Spider orchid 65, *85*
Spruce 159, 175
Staghorn fern 96, *107, 108*
Stanhopea hernandezii 75
Stapelia flavopurpurea 60, *61*
Star magnolia *167*
"Star of the steppe" 36, *46*
Steepwort 103
Stone pine 33, 39, *25, 54*
Strawberry tree 33, *26*
Strelitzia reginae 67, *93*
Strychnos 141, *150*
Suaeda maritima 100
Sundew 103, *120*
Sweet woodruff 135

Tall morning glory 68
Tamarind 14
Tamarindus indica 14
Taxodium distichum 98, 159, *117*
Taxus baccata 175
Telopea speciosissima 38, *49*
Theobroma cacao 11, *4*
Thephrocactus atroviridis 40
Thephrocactus floccosus 40, *15*
Thorn apple 135, 139
Thuja 159
Thymus serpyllum 135

Tien Shan spruce 39, *55*
Tillandsia hildae 105
Tillandsia prodijosa 103
Tillandsia purpurea 16, *23*
Tillandsia recurvata 106, *130*
Tillandsia usneoides 106, *128, 129*
Titanopsis 14
Toothwort 105, 108, *139*
Traveler's tree 36, *44*
Tree heath 33
Trifolium repens 135
Tulip 176, *183*
Tulipa gesneriana 176, *183*
Tulipa sylvestris 33
Tulip tree 11, *3*
Tussilago farfara 68, 139

Umbrella acacia 14
Umbrella pine 158, 159, *162*
Utricularia vulgaris 104

Vaccinium oxycoccos 68, 98
Vaccinium uliginosum 98
Vanilla 95, *101*
Vanilla planifolia 95, *101*
Venus's-flytrap 102, 104, *125*
Verbena officinalis 135
Veriops 99
Vervain 135
Victoria amazonica 11, 174, *175, 176*
Victoria cruciana 174
Viper's bugloss 34, *32*
Viscum album 106, *131, 132*
Viscum laxum 106
Vomiting nut 141

Water hemlock 139
Water lily 11, 174
Water pine 159
Welwitschia mirabilis 12, *6*
Weymouth pine 159
White dryas 39, *52*
White saxaul 16, *24*
Wild tulip 33

Witches' broomstick 135
Witches'-ring 135
Witch-meal 135
Wood sorrel 68

Xanthorrhoea hastilis 38
Xylocarpus 99

Yew 175
Yucca filamentosa 65, *88*

Ahmadjian, Vernon. *The Lichen Symbiosis.* Waltham, Mass.:
  Blaisdell, 1967.

Alexopoulos, Constantine J. *Introductory Mycology.* New York:
  Wiley, 1962.

Bateman, James. *The Orchidaceae of Mexico and Guatemala.*
  New York: Johnson Reprint (reprint of edition of 1837–43).

Benson, Lyman. *Plant Taxonomy: Methods and Principles.*
  New York: Ronald, 1962.

Bold, Harold C. *The Plant Kingdom.* Englewood Cliffs, N.J.:
  Prentice-Hall, 1964.

Boom, B. K., and H. Kleijn. *The Glory of the Tree.* New York:
  Doubleday, 1966.

Bower, F. O. *The Origin of Land Flora.* New York: Hafner,
  1959.

Cady, Leo, and E. R. Rotherham. *Australian Native Orchids in
  Colour.* Rutland, Vt. and Tokyo: Tuttle, 1971.

Cain, Stanley A. *Foundations of Plant Geography.* New York:
  Hafner, 1971.

Cobb, Boughton. *A Field Guide to the Ferns and Their Related
  Families of Northeastern and Central North America.* Boston:
  Houghton Mifflin, 1956.

Core, Earl L. *Plant Taxonomy.* Englewood Cliffs, N.J.:
  Prentice-Hall, 1955.

Corner, E. J. H. *The Natural History of Palms.* Berkeley:
  University of California Press, 1966.

Coulter, Merle C., and Howard J. Dittmer. *The Story of the
  Plant Kingdom.* Chicago: University of Chicago Press, 1964.

Cronquist, Arthur. *Evolution and Classification of Flowering
  Plants.* Boston: Houghton Mifflin, 1968.

Dansereau, Pierre. *Biogeography: An Ecological Perspective.*
  New York: Ronald, 1957.

Daubenmire, Rexford P. *Plant Communities.* New York:
  Harper & Row, 1968.

Dittmer, Howard J. *Phylogeny and Form in the Plant Kingdom.*
  Princeton, N.J.: Van Nostrand, 1964.

Gleason, Henry A., and Arthur Cronquist. *The Natural Geography
  of Plants.* New York: Columbia University Press, 1964.

Good, Ronald. *Features of Evolution in the Flowering Plants.*
  New York: Wiley, 1956.

Good, Ronald. *The Geography of Flowering Plants.* London:
  Longman; New York: Halstead, 1974.

Goodspeed, T. Harper. *Plant Hunters in the Andes.* Berkeley:
  University of California Press, 1961.

Greulach, Victor A., and Joseph E. Adams. *An Introduction to
  Modern Botany.* New York: Wiley, 1962.

Hanson, Herbert C., and E. D. Churchill. *The Plant Community.*
  New York: Reinhold, 1961.

Humboldt, Alexander von. *Aspects of Nature in Different Lands
  and Different Climates.* New York: AMS Press, 1970.

Humboldt, Alexander von. *Personal Narrative of Travels to the
  Equinoctial Regions of America During the Years 1799–1804.*
  New York: AMS Press (reprint of the 1814–29 edition).
  New York: Blom, 1969 (reprint of the 1851 edition).

Hutchins, Ross E. *Plants Without Leaves.* New York: Dodd, Mead,
  1966.

Hylander, Clarence J. *The World of Plant Life.* New York:
  Macmillan, 1956.

Jacobsen, Hermann. *Handbook of Succulent Plants.* New York:
  Humanities, 1973. 3 vols.

Jacobsen, Hermann. *Lexicon of Succulent Plants.* New York:
  Humanities, 1974.

Jamieson, B. G., and J. F. Reynolds. *Tropical Plant Types.*
  Oxford: Pergamon, 1967.

Kramer, Jack. *Orchids: Flowers of Romance and Mystery.*
  New York: Abrams, 1975.

Lamb, Edgar and Brian. *Illustrated Reference on Cacti and
  Other Succulents.* New York: Humanities, 1955–68. 3 vols.

Lloyd, Francis E. *The Carnivorous Plants.* New York: Dover,
  1976.

Luer, Carlyle A. *Native Orchids of the United States and
  Canada.* New York: New York Botanical Society, 1975.

McClure, F. A. *The Bamboos: A Fresh Perspective.* Cambridge,
  Mass.: Harvard University Press, 1966.

Menninger, Edwin A. *Fantastic Trees.* New York: Viking, 1967.

Novak, F. A. *The Pictorial Encyclopedia of Plants and Flowers.*
  New York: Crown, 1966.

Oosting, Harry J. *The Study of Plant Communities.* San Francisco:
  Freeman, 1956.

Padilla, Victoria. *The Bromeliads.* New York: Crown, 1973.

Pears, Nigel V. *Basic Biogeography.* London: Longman, 1977.

Polunin, Nicholas V. *Introduction to Plant Geography.* New York:
  Barnes & Noble, 1967.

Polunin, Oleg. *Flowers of Europe: A Field Guide.* Oxford and
  New York: Oxford University Press, 1969.

Porter, Cedric L. *Taxonomy of Flowering Plants.* San Francisco:
  Freeman, 1967.

Richards, P. W. *The Tropical Rain Forest.* Cambridge and New
  York: Cambridge University Press, 1952.

Riley, Denis R., and Anthony Young. *World Vegetation.* Cambridge
  and New York: Cambridge University Press, 1967.

Sources of illustrations

Scagel, Robert F., and others. *An Evolutionary Survey of the Plant Kingdom.* Belmont, Calif.: Wadsworth, 1965.

Shreve, Forrest, and Ira L. Wiggins. *Vegetation and Flora of the Sonoran Desert.* Stanford, Calif.: Stanford University Press, 1964. 2 vols.

Silverberg, Robert. *Vanishing Giants: The Story of the Sequoias.* New York: Simon & Schuster, 1969.

Sinnott, Edmund W. *Plant Morphogenesis.* New York: McGraw-Hill, 1960.

Tampion, John. *Dangerous Plants.* Newton Abbott: David & Charles; New York: Universe, 1977.

Tivy, Joy. *Biogeography: A Study of Plants in the Ecospace.* London: Longman, 1971.

Van Vijk, W. K., ed. *Physics of Plant Environment.* Amsterdam: North-Holland, 1963.

Wherry, Edgar T. *The Fern Guide: Northeastern and Midland United States and Adjacent Canada.* New York: Doubleday, 1961.

Wherry, Edgar T. *The Southern Fern Guide: Southeastern and South-Midland United States.* New York: Doubleday, 1964.

Withner, C. L. *The Orchids: A Scientific Survey.* New York: Ronald, 1959.

The following persons and institutions kindly supplied the illustrations for this volume:

Archiv Pietasch, Rothenburg (OL): 41
Archiv Streit, Berlin: 39, 42
O. Birnbaum, Halle (Saale): 75
Carolus, Karlsruhe: 143
Conservator of Forest Department, Perth: 50
O. Danesh, Vomp: 27, 28, 29, 30, 54
Dr. F. Ebel, Halle (Saale): 21, 94, 118, 130, 157, 160, 189
Dr. N. Ehler: 105
Dr. G. Gerster, Zurich: 20, 22, 168, 169, 178, 179, 180
E. Hahn, Kirchheimbolanden: 2, 3, 76, 83
K. Herschel, Holzhausen (near Leipzig): 140, 141, 144, 145, 146, 151, 193, 194, 195
E. Hoerler, Stuttgart: 184
R. Höhn, Berlin: 24, 26, 31, 37, 52, 53, 55, 57, 60, 62, 63, 68, 69, 70, 77, 78, 80, 84, 85, 86, 87, 88, 89, 90, 93, 99, 100, 106, 108, 113, 114, 115, 116, 117, 119, 121, 123, 124, 131, 132, 133, 134, 135, 139, 142, 148, 153, 154, 155, 158, 161, 163, 166, 167, 175, 181, 182, 183, 185, 186, 187
Köhler, Berlin: 150
J. Konrad, Leipzig: 95, 190
H. Mathys, Kehrsatz: 71
Naturkundemuseum Berlin: 156
C. Needon, Leipzig: 5, 66, 136, 137, 138, 149, 165
Okapia, Frankfurt on Main: 19, 56, 170
Prof. Dr. W. Rauh, Heidelberg: 4, 11, 15, 16, 23, 32, 33, 34, 43, 44, 45, 46, 51, 58, 61, 64, 65, 72, 73, 74, 82, 102, 103, 109, 110, 120, 122, 125, 126, 128, 129, 152, 162, 171, 172, 188, 192
K. Reis, Frankfurt on Main: 59, 67, 81, 176
W. Richter, Crimmitschau: 17, 91, 96, 97, 98, 104
W. Scheithauer, Bad Aibling: 92
H. J. Schlieben, Pretoria: 7, 9, 12, 35, 36, 38, 40, 101, 107, 111, 112, 159, 173, 174
H. Siebach, Leipzig: 147
Urania-Verlag, Leipzig: 47, 177, 191
Prof. Dr. W. Vent, Berlin: 6, 10
G. Viedt, Berlin: 13, 14, 18, 127
H. Walter, Stuttgart: 8, 25, 48, 49
N. Wisniewski, Berlin: 79
E. Wustmann, Bad Schandau: 1